The Aquaculture Controversy
in Canada

Nathan Young and Ralph Matthews

The Aquaculture Controversy in Canada: Activism, Policy, and Contested Science

UBCPress · Vancouver · Toronto

20 19 18 17 16 15 14 13 12 11 10   5 4 3 2 1

Printed in Canada on FSC-certified ancient-forest-free paper (100% post-consumer recycled) that is processed chlorine- and acid-free.

Library and Archives Canada Cataloguing in Publication

Young, Nathan, 1977-
   The aquaculture controversy in Canada : activism, policy, and contested science / Nathan Young and Ralph Matthews.

Includes bibliographical references and index.
ISBN 978-0-7748-1810-0 (bound); 978-07748-1811-7 (pbk)

   1. Aquaculture – Canada. 2. Aquaculture – Environmental aspects – Canada.
3. Aquaculture – Government policy – Canada. 4. Aquaculture. I. Matthews, Ralph, 1943- II. Title.

SH37.Y69 2010   639.80971   C2010-901125-2

e-book ISBNs: 978-0-7748-1812-4 (pdf); 978-0-7748-5953-0 (epub)

Canadä

UBC Press gratefully acknowledges the financial support for our publishing program of the Government of Canada (through the Canada Book Fund), the Canada Council for the Arts, and the British Columbia Arts Council.

This book has been published with the help of a grant from the Canadian Federation for the Humanities and Social Sciences, through the Aid to Scholarly Publications Programme, using funds provided by the Social Sciences and Humanities Research Council of Canada.

Printed and bound in Canada by Friesens
Set in Stone by Artegraphica Design Co. Ltd.
Copy editor and proofreader: Francis Chow

UBC Press
The University of British Columbia
2029 West Mall
Vancouver, BC V6T 1Z2
www.ubcpress.ca

For my family
— NY

In gratitude to my mentors, whose wisdom,
encouragement, and support set me on my career path:
Ian Whittaker, Roger Krohn, Noel Iverson,
Don Martindale, Reuben Hill, and George Park
— RM

Contents

Figures and Tables

Tables

Acknowledgments

The research reported in this book was conducted with financial support from the Social Sciences and Humanities Research Council of Canada (SSHRC) and the Government of Canada through AquaNet, a Network of Centres of Excellence program.

We gratefully acknowledge the many persons involved in aquaculture issues who gave us their time, knowledge, and advice in conversations, interviews, and questionnaires. We would also like to thank the graduate and postdoctoral research assistants who helped design, collect, and analyze data and findings: (in alphabetical order) Sean Lauer, Mary Liston, Erika Paradis, Gabriella Pechlaner, Amy Poon, Dorothee Schreiber, and Murray Shaw. Special thanks also to Brian Elliott.

We are indebted to three anonymous reviewers and to Randy Schmidt at UBC Press for their comments and suggestions. The usual disclaimers apply.

Abbreviations

AAA	Aboriginal Aquaculture Association
AAC	Aquaculture Association of Canada
AANS	Aquaculture Association of Nova Scotia
AFPR	Atlantic Fisheries Policy Review
AOA	Atlantic Opportunities Agency
ASF	Atlantic Salmon Federation
BCAFC	British Columbia Aboriginal Fisheries Commission
BCLP	British Columbia Liberal Party
BCSFA	British Columbia Salmon Farmers Association
BCSGA	British Columbia Shellfish Growers Association
CAAR	Coastal Alliance for Aquaculture Reform
CAIA	Canadian Aquaculture Industry Alliance
COABC	Certified Organic Associations of British Columbia
DFO	Department of Fisheries and Oceans
DSF	David Suzuki Foundation
EA	environmental assessment
EACSR	External Advisory Committee on Smart Regulation
ENGO	environmental nongovernmental organization
FAO	Food and Agriculture Organization of the United Nations
FDA	First Dollar Alliance
FFAW	Fish, Food and Allied Workers Union
HADD	Harmful Alteration, Disruption, or Destruction
IFOAM	International Federation of Organic Agriculture Movements
NAIA	Newfoundland Aquaculture Industry Association
NBSGA	New Brunswick Salmon Growers Association
NDP	New Democratic Party

NSERC	Natural Sciences and Engineering Research Council of Canada
NSGA	Newfoundland Salmonid Growers Association
OCAD	Office of the Commissioner for Aquaculture Development
PAA	Positive Aquaculture Awareness
PCB	polychlorinated biphenyl
PEIAA	Prince Edward Island Aquaculture Alliance
POP	persistent organic pollutant
POSA	Pacific Organic Seafood Association
RADAC	Regional Aquaculture Development Advisory Committee
SAC	Shellfish Advisory Committee
SAIAC	Salmon Aquaculture Implementation Advisory Committee
SAR	Salmon Aquaculture Review
SCA	Seafood Choices Alliance
SCSA	Special Committee on Sustainable Aquaculture
SDI	Shellfish Development Initiative
SSHRC	Social Sciences and Humanities Research Council of Canada
UBCIC	Union of British Columbia Indian Chiefs
UFAWU	United Fishermen and Allied Workers' Union
WED	Western Economic Diversification
WWF	World Wildlife Fund

Introduction

On the morning of 17 December 2002, members of the Heiltsuk First Nation, along with several environmentalists and commercial fishers, dock fourteen boats at the remote community of Ocean Falls on British Columbia's rugged central coast. The group, numbering nearly sixty, force open the gate to a construction site for a salmon hatchery owned by Omega Salmon, one of the largest aquaculture operators in Canada. Protesters vandalize the site, pulling up the newly poured foundations of the structure. Greg Higgs of the Forest Action Network tells the media: "People are fed up with fish farms. You can expect more protest and more pressure on fish farm companies in the future." Ed Newman, a Heiltsuk elder, is more blunt: "We've declared war on the fish farming industry. They might have to throw a lot of us in jail, but we don't care. We have to protect our way of life."[1]

On 9 January 2004, the internationally renowned journal *Science* publishes an article by University of Indiana chemist Ronald Hites and several colleagues entitled "Global Assessment of Organic Contaminants in Farmed Salmon." The findings include evidence that aquacultured salmon contain significantly higher concentrations of toxic polychlorinated biphenyls (PCBs) than their wild counterparts. The paper's recommendations touch off a maelstrom, as the authors argue that consumers ought to restrict intake of farmed salmon to one serving every two months due to elevated cancer risks. Health Canada, the Canadian Food Inspection Agency, and the United States Food and Drug Administration immediately repudiate the findings, citing the importance of salmon as a source of omega-3 fatty acids, which are essential to the prevention of heart disease. Industry representatives react to the article with venom. In a letter to *Science,* veterinarian and longtime industry booster Brad Hicks writes that "the authors appear to have an agenda beyond the reporting of scientific information ... This is not science. This is politics. This type of report belongs in a political publication where yellow journalism is considered an art."[2]

In the midst of the 2005 provincial election campaign, Carole James, leader of the British Columbia New Democratic Party, is ambushed at a rally in the District of Port Hardy on northern Vancouver Island. Nearly fifty protesters drown out James' speech, chanting, "You'll be in the poor house with the rest of us, Carole!" The hecklers wave signs reading, "Aquaculture Feeds Families" and "We Are Working Environmentalists." They are incensed at the NDP's election pledge to forbid expansion of the salmon farming industry and phase out current technologies. After the incident, James remarks to a journalist: "It's my understanding jobs will be lost anyways, because people are losing their taste for farmed fish."[3]

In August 2005, crab fishers and residents of Chance Harbour on the Bay of Fundy in New Brunswick begin a boycott of aquaculture products. "[Aquaculture] operations are coming to industrialize New Brunswick's shoreline," says an organizer. "This is one of the last remaining stretches of undeveloped coastline west of Saint John. I don't know if there's any more coastline that is as scarce or as spectacular. It's almost sacred."[4]

On 11 October 2005, the Emmy Award–winning ABC television series *Boston Legal* airs an episode in which characters Alan Shore and Denny Crane (played by actors James Spader and William Shatner) vacation in British Columbia, only to discover that wild salmon populations are threatened by sea lice infestations originating from local aquaculture operations. The episode centres on a legal showdown pitting Shore and Crane against industry representatives in a (fictional) courtroom in Port McNeill on Vancouver Island. Mary Ellen Walling, executive director of the BC Salmon Farmers Association, tells the *Vancouver Sun,* "What possible relevance can this have? It's a fictional American television show. William Shatner may be a good actor ... but what possible knowledge could he have of fish biology?"[5]

In 2006, the government of British Columbia strikes a Special Committee on Sustainable Aquaculture to hold public hearings in coastal cities and small communities affected by aquaculture development. The committee hears every imaginable opinion on aquaculture, many of them highly emotional. The committee itself is often strongly rebuked by witnesses. Peter Siwallace, a Hereditary Chief of the Nuxalk Nation states: "I know what recommendations are about. You guys can make recommendations until the cows come in. But would somebody listen to the recommendations? I've got a feeling that the decision is already reached ... I cannot help but think of what happened to the buffalo with the Plains Indians. In order for [settlers] to get the Plains Indians under control, they had to kill off the buffalo. To me, what's happening here is similar. Our fish are being depleted ... You know, there's a reason why we're called First Nations people. We were the first people here on this continent. We were put here by our Creator. You guys were put over in Europe. We accept that fact. You have to accept that fact also." (British Columbia 2006a)

What is going on here? In the past twenty-five years, aquaculture has emerged as one of the most promising but controversial new industries in Canada. Aquaculture, or the farming of aquatic organisms, has been hailed as a potential solution to serious environmental and food supply problems stemming from global overfishing. To critics, however, it has a dark side that poses unacceptable threats to the environment, human health, and local stakeholders. This conflict has mushroomed over the past several decades to become one of the most bitter and stubborn face-offs over industrial development ever witnessed in Canada. The aquaculture industry and its critics are currently locked in a political and cultural struggle that reaches across courtrooms, laboratories, governments, newsrooms, scholarly journals, and virtual and street-level activism.

This is not a book about aquaculture in the conventional sense. It contains only brief discussions of aquaculture techniques, methods, and alternatives. Rather, this is a book about the aquaculture *controversy* – it is an investigation into one of the most divisive and intense struggles over industrial development ever to have taken place in Canada. As such, it is not our intention to reach conclusions about the desirability or sustainability of Canada's aquaculture industry. As we will see, there is no shortage of voices arguing for and against aquaculture development in this country. Rather, we draw on several major research initiatives that we have undertaken over a seven-year period, to present a much-needed analysis of *why* the controversy exists and *how* it is perpetuated.

This book has two aims. First, we look to *analyze* the contours of the aquaculture controversy in Canada – to ask what it is about the collision of interests and claims concerning this industry that is so contentious and divisive. Second, we aim to *explain* the controversy, or at least significant elements of it, by investigating how aquaculture fits in with some of the broader dilemmas and contradictions facing Canada and the world today. We argue that aquaculture has become symbolic of many things to many different people and groups in this country. It has become a metaphor, both positively and negatively, for difficult questions about the collision of humanity and environment, notions of rights and justice, and the rise of intense local/global interactions and conflicts. These are big issues that are increasingly coming to the fore in conflicts over industrial development around the world. Understanding the aquaculture controversy in Canada, therefore, also has relevance beyond this sector and beyond this country. As we will see, one of the reasons why aquaculture is so controversial in Canada is that it is a latecomer – a new claimant to already fragile ecological, economic, and cultural spaces. For this reason, the story of Canadian aquaculture is likely to resonate in many parts of the world dealing with new industries and development in an age of globalization, environmentalism, and the assertion of local rights.

What Is Aquaculture?

At its most basic, aquaculture involves extending the principles of agriculture to marine environments. It typically encompasses the commercial husbandry of fish, plants, and/or shellfish in contained or semi-contained spaces in fresh or ocean waters. For swimming fish or "finfish," this usually involves the suspension of nets in nearshore ocean waters, or more simply the stocking of small lakes and pools for later harvest. Shellfish aquaculture usually involves the seeding of artificial habitat (such as bags or socks that are suspended from pontoons) that is then enclosed as private property.

Aquaculture is an ancient activity, and has been practised for millennia in parts of Africa, Asia, and Europe. Globally, most aquaculture still resembles its ancient form, in which families, communities, and small enterprises stock and feed freshwater species in ponds on a small scale, typically as a supplement to agricultural or fishing activities (FAO 2006). Aquaculture has also recently become big industry and big business, however. Many aquaculture operations are now large-scale, highly mechanized, and owned or backed by some of the world's largest food companies.

The current global controversy over aquaculture is therefore really a controversy over newer, industrial-scale aquaculture. For proponents, aquaculture is simply following the path of terrestrial agriculture, which underwent a "green revolution" in the latter half of the twentieth century, in which family farms gave way in many parts of the world to consolidated or factory-farms that increased productivity from the land by creating economies of scale and investing heavily in irrigation, fertilizers, and pesticides. For supporters of the industry, aquaculture promises a similar "blue revolution" that can harness the full productive potential of marine spaces while reducing pressures on rapidly depleting wild fish stocks (e.g., Economist 2003). For many opponents, however, large-scale aquaculture represents an unjustifiable privatization and industrialization of the oceans. Marine ecosystems are complex and mobile, and cannot be enclosed or isolated from the environmental effects of aquaculture. In opponents' view, high-intensity aquaculture (particularly shrimp and salmon farming) are serious intrusions that may disrupt or destroy ecosystems that are already fragile from years of exploitation and pollution.

There is no mistaking the fact that aquaculture is now a major global industry. According to statistics from the Food and Agriculture Organization of the United Nations (FAO), aquaculture now constitutes roughly one-third of total world fisheries harvest (a total of 52 million tonnes in 2006, nearly double the 27 million tonnes recorded in 1996), and has an annual commercial value of over US$78 billion (FAO 2008, 6). In Canada, aquaculture now dwarfs traditional wild-capture fisheries in some regions of the country. Overall, aquaculture production in Canada has expanded more than threefold from 1991 to 2007 (from 50,000 tonnes to 170,000), although production has recently levelled off due to fierce international competition in aquaculture markets (Statistics Canada 2008). In British Columbia, aquaculture produced 72,000 tonnes of salmon valued at Cdn$364 million in 2007. In comparison, the once-mighty wild-capture fishery landed only 20,000 tonnes of salmon valued at $41 million in the province that same year (British Columbia 2008a).

The Aquaculture Controversy in Canada: An Introduction
Canada is home to both shellfish and finfish aquaculture. These subsectors are structured differently, and are controversial for different reasons. According to some, shellfish aquaculture in Canada can be traced back to Aboriginal harvesting and maintenance of natural clam beds (Tollefson and Scott 2006; British Columbia 2006a). Since the late 1970s, however, shellfish aquaculture has involved the private enclosure of selected nearshore areas for highly controlled seeding, growth, and harvesting, mostly on artificial habitat.

There are several major shellfish aquaculture companies operating in Canada, as well as hundreds of smaller ventures and family businesses. While some Aboriginal groups are directly involved in shellfish aquaculture, the majority of shellfish operations are owned by non-Aboriginal firms and entrepreneurs. The shellfish sector is also relatively small, accounting for only 10 percent of the total value of Canadian aquaculture.

Finfish aquaculture in Canada involves both freshwater and saltwater species. The most common freshwater species is rainbow trout, which is typically raised in ponds on private property, or else in cages or nets that are suspended in lakes (Michalska 2005). Like shellfish aquaculture, most trout-farming operations are small and family-owned. Salmon aquaculture is the largest and most lucrative area of Canadian aquaculture, representing 75 percent of the industry's total value. While salmon aquaculture in Canada was once dominated by small businesses, it is now the near-exclusive province of large firms, many of which have multinational operations. Salmon aquaculture is also the most controversial subsector of the industry. While criticisms of shellfish and freshwater aquaculture are becoming more common (e.g., Deal 2005; Michalska 2005), salmon aquaculture is the primary target of activism. Environmentalist groups and other opponents routinely accuse the salmon aquaculture industry of causing serious environmental damage, threatening human health, and violating stakeholder rights. Supporters of salmon aquaculture have countered these allegations primarily by arguing that the industry has a minimal environmental footprint and is a major source of employment and stability in economically distressed coastal communities (e.g., Bastien 2004).

Four Main Axes of Controversy
The controversy over aquaculture in Canada is complex and multidimensional, involving competing claims from many different actors. As we will discuss in Chapters 1 to 3, environmental and industrial conflicts have become more complex in recent years because of greater public sensitivity to issues of risk and uncertainty, the diffusion of knowledge- and information-generating capacities to non-traditional actors (including environmentalist and other activist groups), and greater political sensitivity to stakeholder and citizen rights (Irwin 1995; Fischer 2000). Governments and industries, which for a long time enjoyed the final word on development and risk management, are still learning how to deal with such complexities. Leiss (2001) uses the apt metaphor of a labyrinth to describe the volatility that results from the involvement of activists and an empowered public in issues that were previously settled in boardrooms and government offices. As we will see throughout this book, the aquaculture controversy is indeed labyrinthine, as new claims and information about aquaculture are constantly

Figure 0.1

Major axes of the aquaculture controversy in Canada

Axis	Key points of contention
Environment	Does industry belong in marine environments?
	Does aquaculture threaten or benefit wild fish stocks and other aquatic species?
Human health	Is aquaculture beneficial or harmful to human health?
Rights	Is it acceptable to enclose "public" marine spaces for private gain?
	Does aquaculture disrupt local and/or Aboriginal rights and claims?
Rural development	Is aquaculture beneficial or harmful to rural development and well-being?

emerging and interacting with people's hopes, fears, and ideas about fairness and justice.

Although the aquaculture controversy is complex, patterns can be observed in the conflict. The aquaculture labyrinth in Canada is composed of four main axes or dimensions: the environment, human health, rights, and rural development (Young and Matthews 2007a). As mentioned, these apply most intensely to salmon aquaculture, but also to shellfish and freshwater aquaculture. Each axis represents an identifiable master theme in the aquaculture debate, although the lines among these are often blurred in real-world conflicts and narratives. Figure 1 summarizes the main points of contention between supporters and opponents of aquaculture in Canada across these dimensions.

Environment
Like all resource industries, aquaculture intrudes on the natural environment. Supporters argue that the industry is a legitimate user of marine spaces that should have equal rights and access to the oceans with commercial fishers and other traditional users (e.g., Bastien 2004). Proponents also assert that the industry has tangible ecological benefits. World fish consumption is increasing every year, and supporters argue that aquaculture reduces pressure on overexploited wild fish stocks (e.g., Dhaliwal 2000). Many opponents, however, see high-intensity aquaculture as an unjustifiable industrialization of nature. First, critics often argue that rearing carnivorous salmon in fact depletes ocean resources, as aquaculture prompts "fishing down the food

web," where smaller species of fish are harvested in order to provide food meal for the aquaculture industry (cf. Pauly et al. 2000). Opponents have also long complained that salmon aquaculture sites are major polluters through the release of fecal matter and uneaten feed into the marine environment (Ellis 1996). They also frequently point to the problem of escapes. The vast majority of salmon grown in Canada are Atlantic salmon (*Salmo salar*), which are preferred to other species because of their hardiness and faster growth rates. Occasionally, large numbers of these captive fish escape due to human or mechanical error, damage to the nets that contain the fish, or violent weather. This raises a number of concerns on both Atlantic and Pacific coasts. In the Atlantic region, native populations of Atlantic salmon are critically endangered, and there are fears that escaped salmon may compete with fragile wild stocks, or else genetically alter them through interbreeding (ASF n.d.). In the Pacific region, Atlantic salmon are an exotic species, and critics fear that escaped fish may be colonizing Pacific waterways (e.g., Volpe 2001).

Another major environmental issue involves the transfer of diseases and parasites. Salmon farms have very high population densities, which facilitates the transfer of pathogens among captive fish. The real point of contention, however, is whether this has an effect on wild species. A particularly damaging allegation is that aquaculture sites are transferring parasites called sea lice (*Lepeophtheirus salmonis* and *Caligus clemensi*) to migrating wild juvenile salmon (e.g., Morton et al. 2001). The sea louse is a naturally occurring ocean parasite, but multiple lice infections can be fatal to juvenile salmon (the precise mortality rate remains a point of contention). Some critics have blamed sea lice infections for unexpectedly low returns of wild salmon in recent years to some British Columbia rivers (Morton 2004; Krkosek 2005; Krkosek et al. 2006), although industry supporters strongly dispute this conclusion (Brooks 2006; Butterworth et al. 2006).

Human Health

The salmon aquaculture industry promotes its products as being "farmed fresh and healthy" (PAA 2008), and points to multiple studies demonstrating the health benefits of consuming salmon of all kinds (due mostly to high levels of omega-3 and omega-6 fatty acids). Some opposition groups, however, have raised objections about the use of antibiotics, pesticides, and artificial colourants in salmon aquaculture production as both a threat to worker safety and to public health (e.g., DSF 2007). Moreover, a series of studies (some of which were funded by environmentalist organizations) have claimed that farmed salmon contain elevated levels of PCBs, dioxins, and other persistent organic pollutants (POPs) that are transmitted through feed (Easton and Lusniak 2002; Jacobs et al. 2002a and 2002b; Hites et al. 2004). Hites and colleagues' article "Global Assessment of Organic Contaminants in Farmed Salmon" appeared in the journal *Science* and was carried by media outlets worldwide in 2004, leading to a significant drop in farmed salmon sales across Europe and North America (Leiss and Nicol 2006). In short, the competing claims regarding the health risks and benefits of aquaculture products put consumers in a difficult position, asking them to weigh one malignancy (cancer) against another (heart disease).

Rights

The aquaculture industry has elbowed its way into Canada's marine spaces and established islands of private property in the ocean "commons." There is a longstanding tension in Canada between the drive to centrally administer marine waters in the national interest and the notion that marine resources belong, if they belong to anyone, to coastal users and communities (cf. Matthews 1993; Phyne 1994). Commercial aquaculture treads directly on this sensitivity.

First, shellfish aquaculture takes place almost exclusively in beach and nearshore spaces, leading to conflicts with property owners, traditional users, tourism operators, and some First Nation groups who have rights claims to these areas (Howlett and Rayner 2004). Second, salmon aquaculture in particular raises the spectre of distant powers interfering in coastal culture and livelihoods. The early pioneers of salmon aquaculture in Canada were entrepreneurs who proceeded largely by trial and error, often learning the basics of aquaculture from pamphlets and tacit knowledge derived from enhancement projects or commercial fishing (Keller and Leslie 1996, 13). From the 1980s onward, however, the Canadian salmon aquaculture industry has progressively consolidated and is now dominated by a handful of large firms, most of which are based outside of Canada. Historically speaking, rural communities have reason to be wary of big companies and big industries. One of Canada's great economic advantages has been its vast and varied resource wealth, which has enabled it to become highly prosperous despite a limited manufacturing and financial base (Innis 1956). Resource exploitation has taken a toll on many communities in the Canadian hinterland, however. The forestry, mining, and energy industries have legacies of carelessly degrading local environments with the full blessing of government. For some, the aquaculture industry looks suspiciously like the latest in a long line of exploiters seeking to profit from permanent damage to local environments and livelihoods.

Finally, aquaculture has landed squarely on the shoulders of the Aboriginal rights movement. Aboriginal rights to marine spaces and resources are hotly contested on both Atlantic and Pacific coasts, and many aquaculture operations are found in spaces that are claimed as traditional territories by one or more Aboriginal groups. Rights over traditional territories are still unclear, although several recent rulings in Canadian courts have confirmed their legitimacy (Woolford 2005). As we will see in Chapter 2, the issue of Aboriginal rights is rather complex, and is mobilized both in opposition to and in support of aquaculture development. Nevertheless, the Aboriginal rights movement has gained strong political and legal momentum, and has greatly empowered individual First Nations communities where questions of local and regional resource management are concerned.

Rural Development

The final axis of the controversy outlined in Figure 1 involves the question of rural development. Canada's coastal regions have experienced great economic hardship since the early 1990s due to declines in traditional industries such as fisheries and forestry. These changes have profoundly affected social and economic structures in coastal communities. Research by Sinclair and colleagues (1999) in Newfoundland following the closure of the northern cod fishery found deep despondency and social disruption in affected communities. The sudden collapse of the fishery did no less than "wrench away the underpinning of local social structures" by disembedding families from the community practices "that had allowed people to maintain their lives through combinations of personal subsistence labour, paid work, and state aid" (Sinclair et al. 1999, 327). On the Pacific coast, declines in fisheries and forestry have thrown many communities into similar disarray as traditional patterns of work, spending, and social status tied to these industries recede (Young 2006a).

But while traditional industries are down, they are not out. As we will see in Chapter 2, many coastal residents are still deeply committed to traditional activities and industries. Supporters of aquaculture frequently argue that the industry is a much-needed economic bright spot in coastal regions, capable of countering job losses in other sectors (e.g., Shumway et al. 2003; Bastien 2004). Some local critics, however, instead see aquaculture as an unacceptable threat – even a final straw – for already vulnerable commercial and subsistence fisheries. While the causes of the decline of traditional fisheries are complex (including domestic overfishing, habitat destruction from other industries such as forestry, and international jurisdiction disputes on both Atlantic and Pacific coasts), aquaculture has become a key expression point for grievances over the hardships that have resulted from this deep loss. Indeed, we will argue later that aquaculture has come to be for many people a metaphor for the failures of government and industry to preserve coastal resources and ways of life.

The Players

The cast of characters in the aquaculture controversy in Canada is long and varied, and includes governments, activists, scientists, consumers, and local stakeholders. The clearest division among major players in the controversy, however, is between groups with a mandate to promote and defend the industry and groups that have adopted a clearly oppositional stance. The aquaculture industry in Canada is served by a large number of associations. On the Pacific Coast, major associations include the British Columbia Salmon Farmers Association (BCSFA), the British Columbia Shellfish Growers Association (BCSGA), and the Aboriginal Aquaculture Association (AAA). Industry associations on the Atlantic coast include the New Brunswick Salmon Growers Association (NBSGA), the Newfoundland Aquaculture Industry Association (NAIA), the Newfoundland Salmonid Growers Association (NSGA), the Aquaculture Association of Nova Scotia (AANS), and the Prince Edward Island Aquaculture Alliance (PEIAA). At the national level, the industry is represented by the Canadian Aquaculture Industry Alliance (CAIA), and the Aquaculture Association of Canada (AAC).

These associations have been instrumental in lobbying provincial and federal governments and in serving as public advocates and media voices for the industry (as we will discuss in Chapter 5). CAIA, the BCSFA, and the NBSGA have been the most active public defenders of the industry, often engaging directly in protracted struggles with opposition groups over the production and interpretation of claims about aquaculture. There are also several organizations representing pro-aquaculture stakeholders that have recently entered the fray as defenders of the industry, including Positive Aquaculture Awareness (PAA) and the First Dollar Alliance (FDA).

Opposition groups are equally diverse and involve alliances across local, regional, national, and international scales. Most opposition groups are self-identified as environmentalist nongovernmental organizations (ENGOs). On the Pacific coast, ENGOs active in the aquaculture controversy include the David Suzuki Foundation, the T. Buck Suzuki Foundation, the Georgia Straight Alliance, the Living Oceans Society, the Raincoast Conservation Society, the Friends of Clayoquot Sound, and the Watershed Watch Salmon Society. Each of these groups is a member of the Coastal Alliance for Aquaculture Reform (CAAR), which has served as a collective means of conducting research and communications. CAAR's main campaign, "Farmed and Dangerous," targets salmon aquaculture and has been notably successful in organizing boycott initiatives and other means of pressuring retailers and restaurants to declare themselves "proud to be farmed salmon free." There are significantly fewer groups on the Atlantic coast and they are generally less obstructionist. Perhaps the most active has been the Atlantic Salmon Federation (ASF), which acknowledges that the industry has reduced pressure on endangered wild stocks but at the same time argues that the industry presents significant threats through pollution and escapes (ASF n.d.).

These local and regional groups are the main actors in the Canadian controversy, but they have also been joined by some international ENGO heavyweights. For instance, the US-based Pew Charitable Trusts support research into the environmental and health effects of aquaculture. The Sierra Club of Canada has taken the government of Nova Scotia to federal court over the expansion of shellfish farming along the province's famed Cabot Trail. Greenpeace Canada has petitioned the Auditor General of Canada to investigate the policies of the federal Department of Fisheries and Oceans (DFO) regarding the use of genetically modified fish in aquaculture. The World Wildlife Fund has a strong international campaign to reform finfish aquaculture and enhance marine protection, and has partnered with the ASF in key research and lobbying activities (cf. Porter 2003).

While ENGOs are often the most visible, other groups are involved in the opposition movement. These include Aboriginal organizations such as the Union of British Columbia Indian Chiefs (UBCIC) and the British Columbia Aboriginal Fisheries Commission (BCAFC). Both organizations use very strong language in their opposition to salmon farming in particular, and the UBCIC has held a zero-tolerance position on salmon aquaculture since 1998 (UBCIC 1998). Some commercial fishers' organizations have also taken strong oppositional stances. The United Fishermen and Allied Workers' Union (UFAWU) in British Columbia has expressed opposition to aquaculture primarily through its affiliation with the ENGO the T. Buck Suzuki Foundation, but the union has also taken a direct stance against the industry, with President John Radosevic being on record saying that "the BC government is willing to sell out ten thousand commercial fishermen and fish plant workers for the sake of a handful of multinational corporations" (Simpson and Beatty 2002). On the Atlantic coast, the powerful Fish, Food and Allied Workers Union (FFAW) has targeted aquaculture as part of a broader campaign against DFO's controversial Atlantic Fisheries Policy Review (AFPR). The AFPR was initiated under the Liberal government of Jean Chrétien in 1999 to reform fishery policy in light of the permanent extension of the moratorium on the commercial northern cod fishery. As such, it has placed a strong emphasis on conservation and diversification of ocean industries, including aquaculture (DFO 2004). The FFAW's particular objection is that "after years of struggle and a commitment to rebuild stocks, fishing people are now being asked to accept new users when they have yet to recover the ground they lost and when stocks have yet to recover anywhere close to levels where it is safe or sustainable to introduce new users" (FFAW 2001).

These are the main players in the aquaculture controversy in Canada. As we will see in Chapter 3, these groups are engaged in intense struggles to sway public opinion. As in many political conflicts, this involves both positive and negative claims making. On the positive side, both pro-aquaculture and opposition groups argue that they are acting in the best interests of the environment and of coastal stakeholders (often to the frustration of actual stakeholders – see Schreiber and Newell 2006a). On the negative side, these groups also engage in significant efforts to discredit their adversaries.

This often involves direct trading of claims and counterclaims that aim to expose opponents' agendas and biases (Hume et al. 2004). These claims often make it into the media, which contributes to the climate of hostility and distrust on all sides of the issue (see Chapter 5). Aquaculture disputes are also clogging the courts on both coasts, with multiple suits pending against government and industry over aquaculture practices and siting (Riley 2003; Sierra Legal Defence 2003; Atkinson 2007). As we will see throughout this book, the battle over the industry's public image takes place on multiple fronts.

A Sociological Analysis of the Aquaculture Controversy
The aim of this book is to advance a sociological analysis of the aquaculture controversy and its broader implications for all kinds of new industrial development. Generally speaking, sociology is the study of patterns in human relationships and other types of formal and informal social organization. This perspective prompts us to look at the aquaculture controversy in a particular way. Aquaculture, and especially salmon aquaculture, is a very science-heavy activity. Natural scientists and engineers in Canada and around the world are making great progress in better understanding and managing the effects of aquaculture on the natural environment. Despite these advances, however, the controversy continues to intensify rather than abate. To paraphrase Ulrich Beck (1992), this is a case where more and more knowledge paradoxically appears to be leading us further and further from consensus. In our view, sociology can make a real contribution by taking a broader view of *how* and *why* the conflict is perpetuated as a political and cultural phenomenon, rather than as a question for the natural sciences alone.

The analysis that we present in this book is based on several years of in-depth research into multiple dimensions of the aquaculture controversy. Our work has involved survey research, field studies, and analysis of public artefacts such as media coverage, interest group publications, and the transcripts of public meetings and hearings on aquaculture.

We conducted four separate survey projects in the course of our research. One was a survey of experts in aquaculture in Canada across all sides of the debate, including experts in academia, government, industry, and environmentalist groups. We also conducted a large survey of aquaculture firms operating in Canada, a smaller survey of aquaculture workers in coastal regions, and a survey of local businesses in one aquaculture-intensive community (Port Hardy, British Columbia). This research gives us a means of analyzing some of the most contentious issues in the aquaculture controversy. Importantly, it also provides an opportunity to hear directly from the people who are involved in aquaculture issues but whose voices are not always discernible in the public clashes between organized pro- and anti-aquaculture groups.

The thesis of this book is that the aquaculture controversy is both unique (reflecting the specific history of coastal and resource development in Canada) and rooted in some of the major unresolved questions about environment, development, rights, and governance confronting democratic societies around the world. This analysis is made in three main parts.

Part 1 (Chapters 1 and 2) develops the argument that the aquaculture industry and controversy in Canada are being shaped by the intersection of multiple economic, political, and cultural developments that are unfolding both in coastal regions and in the broader Canadian and global society. Aquaculture is a young industry, and we suggest that this means two things. On the one hand, aquaculture is being exposed in a formative way to some of the major economic and political forces to emerge in the past two decades. Canada's traditional resource industries, such as forestry, mining, and capture fisheries, have each been fundamentally changed by developments such as economic globalization, the environmental movement, and the Aboriginal rights movement (cf. Wallace 2002; Hayter 2003). The difference with aquaculture, however, is that it is encountering these forces while still in its infancy. On the other hand, we will argue that the late arrival of aquaculture means that the industry has *come to represent these issues* to many stakeholders. In many respects, aquaculture has become a line in the sand for activism *about* globalization, *about* environmentalism, and *about* Aboriginal rights. In a sense, the industry itself has been guilty of inviting these associations. We argue that by claiming that aquaculture can save coastal communities suffering from economic recession and alleviate environmental problems (such as overfishing), the industry itself opened the door for these themes to enter the debate. More seriously, by casting itself as an economic and environmental problem solver, the industry may have married itself *to the problem.* This association with some of the most urgent challenges facing coastal Canada leads us to argue that *aquaculture has become a metaphor for a myriad of hopes and fears facing coastal stakeholders and concerned citizens elsewhere in the country.* This is ultimately a nightmarish scenario for the industry. We argue that although aquaculture is still a relatively minor player in Canada's resource economy, it has become a flashpoint for complicated hopes and fears about past and future. For many of its supporters, the industry is a way forward and a way to salvage a rural coastal way of life. For many critics, it represents the end of these things. It is difficult to find compromise in these visions.

Part 2 (Chapters 3, 4, and 5) examines knowledge and communications conflicts about aquaculture. Both pro- and anti-aquaculture groups are engaged in the production and dissemination of competing claims about the industry. Many of these claims draw heavily on scientific research and, more broadly, the authority that science wields in Western culture as a disinterested and privileged type of knowledge. In Chapter 3, we look at different ways in which science, narrative (or "framing"), and communications strategies are mobilized by activists on all sides of the debate to try to influence public opinion.

In Chapter 4, we look at data from the survey of aquaculture experts in Canada to examine how individual scientists understand the controversy and their own roles in it. The survey demonstrates clearly that there is no consensus on the effects of aquaculture, but we find that opinions on aquaculture are highly patterned. Experts' opinions are significantly associated with variables such as institutional affiliation, career history, professional networks, and personal values. This suggests that differences of opinion among scientists run much deeper than simple disagreements over facts and interpretation. Findings from the survey also indicate that the aquaculture controversy is having a corrosive effect on the scientific community working in this field. Science is a cumulative effort that depends on the ability of scientists to trust one another's findings. Many respondents expressed deep mistrust of the motives and methods of those who disagree with them, and this does not bode well for future consensus or compromise on aquaculture. Last, we look at experts' views of stakeholders and the general public. Findings are mixed, as we see that experts from all sides of the debate are open to public involvement in aquaculture regulation and decision making but are also critical of the public's ability to arrive at responsible conclusions about aquaculture.

Chapter 5 examines media coverage of aquaculture in Canada. Both supporters and opponents of aquaculture regularly complain that media coverage is biased against their position. Our study of aquaculture coverage in seven daily newspapers in Canada in fact shows a relative balance in coverage of pro- and anti-aquaculture views. On closer analysis, however, we find that media coverage is highly selective – focusing on particular voices and thematic "packages" that tell particular types of stories about the industry and controversy. Among our most significant findings is that *industry voices* appear most frequently in media coverage (they are quoted more often than any other group), but that *oppositional themes*, particularly regarding environmental risks and harms, are much more common than pro-aquaculture themes. This suggests that while industry voices appear to be dominant, they are often being quoted regarding the problems of aquaculture rather than its benefits.

Part 3 (Chapters 6 and 7) examines the political economy of Canadian aquaculture. Chapter 6 looks at the effects of aquaculture on rural economies. Using our survey of aquaculture firms in Canada, we address the contentious issues of employment and job quality. There are conflicting claims regarding the number of jobs in Canadian aquaculture. Our own estimates based on the survey come in at the lower end of the spectrum (between 5,000 and 6,000 full-time, part-time, and seasonal jobs Canada-wide). We also find evidence, however, that many aquaculture jobs, particularly in salmon aquaculture, are relatively high paying and stable. This conclusion is bolstered by the (limited) companion survey of aquaculture workers, which finds high levels of job satisfaction among salmon aquaculture employees. Chapter 6 also examines the relationship between aquaculture firms and local economies. Over the past two decades, Canada's coastal communities have suffered substantial economic hardships. Aquaculture promises a way to compensate for declines in traditional resource sectors – a means to re-establish community stability and provide employment for young people who wish to stay or return home. At the same time, critics argue that aquaculture harms existing businesses that are already vulnerable due to the economic downturn (particularly in tourism and other activities that depend on pristine environments). Our research finds strong evidence that aquaculture firms make significant use of local services through direct contracts and other, informal means. Our study of the business community in Port Hardy also finds that few businesses feel that they have been harmed by the aquaculture industry.

Chapter 7 addresses contentious issues surrounding the governance of aquaculture in Canada. Opponents of aquaculture express serious reservations about the dual role that has been adopted by government as both promoter and regulator of the industry, while supporters argue that the industry is in fact over-regulated because of jurisdictional overlaps and political sensitivities. These competing claims restrict the range of options open to policy makers and regulators. We argue, however, that the main challenge facing governments comes from three urgent pressures that are pulling aquaculture policy in opposing directions. The first pressure is economic. The aquaculture industry in Canada is relatively small, and is having a difficult time competing in the international market for aquaculture products (we discuss this in detail in Chapter 1). In this competitive environment, "cumbersome" regulations are seen as a dangerous drag on productivity and flexibility in the global market. The second pressure is environmental. Aquaculture clearly has an environmental footprint that must be regulated by the state. At the same time, government policy makers are not ignorant of the controversy, and develop environmental policy in dialogue with (and as a means of rebutting) oppositional narratives. The third pressure has to do with the legitimacy crisis facing the industry. Governments are keenly aware that aquaculture will not survive without broad stakeholder and public acceptance (DFO [2000] has made "building public confidence in the industry" a policy priority).

In order to deal with this triple pressure, federal and provincial governments have mobilized both traditional and innovative policy responses. On the traditional side, governments have created numerous generous subsidy programs to reduce industry costs and thus enhance competitiveness.

Alongside these, however, governments have also moved (in limited fashion) to implement principles of what has been called "smart regulation," "results-based regulation," or "self-regulation," which put greater onus on the industry to participate in aquaculture policy development and to monitor environmental compliance. While this approach is controversial, it is understood by governments as a means of simultaneously enhancing flexibility and improving environmental performance. At the same time, governments are also engaging in legitimacy-building exercises (with mixed results) that try to include stakeholders and the general public in aquaculture development.

Our concluding chapter argues that the aquaculture controversy in Canada is about more than it seems. Aquaculture has become a make-or-break issue for many interests on all sides of the debate. While the debate itself is often highly technical (about the existence and magnitude of risks, pollution thresholds, ecosystem interactions, and so on), we argue that the controversy endures largely because it has come to mean so many different things to different people. This complexity means that the aquaculture controversy will not be resolved through bullying, shouting, or efforts to simplify the debate. Returning to Leiss' term (2001), the controversy is similar to an enormous labyrinth, where players hold long-term visions and goals for the future of aquaculture but are forced into immediate actions and reactions that lead in unpredictable directions. In what follows, we attempt to find patterns in the labyrinth – to explore and explain the contours of this stubborn controversy.

Part 1
A High-Speed Collision: Aquaculture as Intersection and Metaphor

Large-scale commercial aquaculture is the first genuinely new resource industry in Canada in a long while. This country has a long history of agriculture, forestry, fisheries, and mining. Even more recent sectors, such as fossil fuel extraction, are often considered offshoots of older, familiar industries such as mining. We will argue that the late emergence of aquaculture is important in several respects for understanding the current controversy. First, the industry's late arrival has meant that public understanding of aquaculture began with a *tabula rasa*. In the early days of the industry, aquaculture was frequently portrayed as an extension of traditional commercial fisheries (Robson 2006, 35). This began to change in the late 1980s with the emergence of new environmental issues – such as pollution and disease incubation – that had not been encountered with traditional fisheries (Keller and Leslie 1996). We argue that because of this novelty, the "blank slate" has ultimately been a significant liability for the industry. As a new activity with unfamiliar methods, aquaculture has been uniquely vulnerable to the efforts of opposition groups to fill in the blanks in public and stakeholder understandings of the industry. Second, the late emergence of aquaculture means that both the industry and opposition movements are being formed in the crucible of major contemporary economic, political, and cultural movements. Whereas traditional resource industries in Canada were shaped by a political economy of centralized authority over resource management, high-volume mass production, and the rigid structuring of access rights to resources, aquaculture has grown up in a very different set of contexts.

In the following two chapters, we will argue that these contexts are both local and global. In Chapter 1, we posit that the aquaculture industry and controversy in Canada are being strongly influenced by global-scale forces such as economic globalization, the rise to dominance of neoliberal perspectives and methods of governance, and the growing influence of the global environmentalist movement. In Chapter 2, we will discuss how the aquaculture industry and controversy are also being shaped by specific local and regional challenges, such as the severe economic downturn in coastal Canada, as well as the increasing legal, political, and moral weight behind the Aboriginal rights movement in this country.

The primary argument advanced here is that the debate over aquaculture in Canada is about much more than aquaculture alone. In our analysis, aquaculture has become both an *intersection* and a *metaphor.* We suggest that aquaculture is a locus where multiple global and local tensions are meeting and expressing themselves. The aquaculture industry (whether it likes it or not) has become a site where much broader pressures and struggles collide. At the same time, we will argue that aquaculture has come to be a powerful but very flexible and fluid metaphor, particularly in local settings. As a new industry arriving at a time of change, uncertainty, and unease, aquaculture has become a metaphor for both hopes and fears, dreams and nightmares. For some stakeholders, aquaculture has come to represent the best of past, present, and future; for others, it is the worst. As we will see, this fluidity of meaning makes the debate over aquaculture highly complex and deeply emotional.

1

Aquaculture in a Global Context

The aquaculture industry in Canada is emerging at a time when the traditional logics of industrial and resource development are being challenged on multiple fronts. In this chapter, we argue that these challenges stem from important changes in global capitalism, politics, and civil society. While Canada's resource economies have long been dependent on exports and foreign investment, the maturation of a truly global capitalist marketplace is significantly complicating Canada's once-dominant resource and agriculture sectors (Klein and Kerr 1995; Hayter 2003). As we shall see, global capitalism exerts complex pressures on primary economies in this country, sometimes providing impetus to growth and at other times (and places) impetus to restructuring and decline (Young 2008). Moreover, these economic pressures are colliding with the limits to growth imposed by environmental exhaustion. Traditional forestry and fisheries have struggled for several decades with severe problems stemming from environmental degradation (Marchak et al. 1999; Ehrenfeld 2005). In combination, these forces are threatening Canada's longstanding (and perhaps dubious) status as a world leader in high-volume resource production (Burda and Gale 1998; Hayter 2000).

The challenges from politics and civil society are equally complex. Although globalization has made certain corporate actors more powerful and free, other changes may be restricting these freedoms. As we shall see, organized environmentalism has become a momentous global force. More generally, over the past several decades, the notions of stakeholder and consumer rights have gained real legal and moral traction in democratic societies. For a long time, government and industry could count on their combined weight to push through controversial projects or actions. Although this is still true, businesses are now more vulnerable to citizen protest than in the past (Fischhoff 1995). This is due in part to new legal obligations, such as the requirement (following the 1997 *Delgamuukw* decision of the Supreme Court of

Canada[1]) that firms and governments consult with Aboriginal stakeholders before proceeding with projects that may encroach on group rights. But it is also due to cultural shifts in the relationship between businesses and the general public, which has become much less tolerant of corporate malfeasance. The business management literature now recognizes many cases, including British Columbia's "war in the woods" and the European Union's restrictions on genetically modified foods, where large companies have been broken by their failure to recognize the power of citizen movements to paralyze production and/or throw up impassable political roadblocks (Magnusson and Shaw 2002; Leiss 2001).

Even as the economic and cultural rules of the industrial game are changing, so too are government involvement and oversight of resource development. Federal and provincial governments in Canada have traditionally seen resource development as an opportunity for nation and province building (Matthews 1983). Older models that coupled resource development with welfare-state programs, high employment, and centralized regulatory regimes no longer anchor government approaches to resource development, however. In place of these older strategies, governments are now turning to more flexible, nimble, and "smart" forms of resource policy, both in the name of allowing greater efficiencies in resource production and in the name of improving environmental compliance (Young and Matthews 2007b).

All of these changes are controversial in their own right, and are contested at global, national, and local scales. Moreover, although these trends are concurrent and linked in many respects, they are also often contradictory. In this chapter and the next, we argue that the aquaculture industry is caught in the intersections of these forces – pressured by the logic of global capitalism to expand and grow, to pursue efficiencies, and to take risks, and pressured by stakeholders, consumers, and activists to constantly justify its methods, products, and rights to marine spaces.

Globalization and Natural Resources

Social scientists are divided about what is actually new about economic globalization. Some point out that international trade and interdependency have been a reality since at least the nineteenth century (e.g., Hirst and Thompson 1996), while others argue that current economic globalization is fundamentally rewriting the economic and political rules of the game (e.g., Castells 2000; Urry 2007). We argue that current economic globalization is having a demonstrable impact on Canadian aquaculture in two ways. First, the aquaculture industry in Canada is subject to the pressures of increasingly globalized markets (particularly in terms of cost and price). Canadian-based aquaculture producers hold a small share of the global market, which makes them rather vulnerable to competition and price fluctuations. Second, globalization has prompted deep changes in traditional resource industries such as commercial fisheries and forestry. These changes have led to severe reductions in employment and investment in coastal Canada. Aquaculture has emerged in the midst of these changes – which has been both good and bad for the industry. As we will discuss later in this chapter, at times the aquaculture industry has actively promoted itself as a saviour for communities and workers who have fallen victim to declines in traditional industries. Sometimes, however, the global character of the industry is a liability in the eyes of local people seeking to avoid the mistakes of the past.

Resource industries and economies have long been international in orientation. Indeed, much of the early scholarly work on the logic and structure of global economies was based on studies of resource industries. For instance, the theory of comparative advantage developed by classical political economists such as David Ricardo advocated freer international trade so that nations could concentrate on producing certain goods that reflected environmental and/or labour advantages, while importing goods that were more advantageously produced in other nations (cf. Ricardo 1969 [1817]). This became the accepted wisdom of the nineteenth century, and colonial territories and rural areas were encouraged to develop industries based on resource wealth (cf. Goldin 1990). The link between international resource economies and local development has been particularly important for Canada, as both a former British colony and neighbour to the world's most industrialized nation. Indeed, it was the Canadian economic historian Harold Innis who developed a strong counterpoint to theories of comparative advantage in his "staples thesis" (Innis 1933).

The staples thesis is both an economic and an institutional theory. It accepts the main argument of the theory of comparative advantage (that regions will specialize in any geographically substantial commercial space), but it argues that this is not always to the advantage of peripheral regions. In an environment where investors are free to move their capital where it is best suited, investment in peripheries will target development of single activities and industry to feed better-established manufacturing and service centres located elsewhere (Marchak 1983, 23). In other words, this investment makes strong demands on local institutions (shaping labour markets, local economies, and even local governments) without committing in any way to local diversification or long-term development. Thus, the staples perspective introduced an important power dimension to understandings of international trade and movement in commodities, suggesting that peripheral regions are actively shaped by the market demands and political influence of the centre. Among the key observations of staples theory is that peripheral resource regions bear many of the economic and environmental risks necessary for the stability of central economies (Marchak 1983, 23). While the theory of comparative advantage implies that resource regions ought to be capable of leveraging their natural wealth towards long-term development, the staples perspective emphasizes the vulnerability of peripheries to dependency on forces well beyond local control.

Given that Canada's resource industries have long been oriented towards export and international markets, it is important to return to the question of what is new for resource peripheries in the current round of economic globalization. For instance, some scholars continue to place staples theory at the centre of their analyses of restructuring in Canada's resource sectors (e.g., Markey et al. 2005), which suggests a degree of continuity with prior experiences. At the same time, however, other social scientists argue that resource industries and regions are being fundamentally reshaped in this latest round of economic globalization (e.g., Hayter 2003; Young 2006a).

Research suggests that both are probably correct. Taking the broad view, what is new about the current experience of globalization is its completeness. For the first time in human history, we can talk about truly global capitalist markets for commodities and resource products. To be clear, borders and distance are not going away. For evidence of this, we need only look to Canada's trade relationship with the United States (a country that received 77 percent of Canadian forest products exports in 2002, at the height of the bitter softwood lumber trade dispute). Expanding economic globalization has opened the door, however, for new producing regions to challenge Canada's traditional dominance in resource exports. Compounding this change is the fact that Canadian resource sectors have traditionally focused on high-volume, low-value production. Scholars have argued that Canada's vast wealth of resources, coupled with a dependency on export to world centres, has led the nation into a "staples trap," where high-volume production has trumped the development of secondary manufacturing and research and development (cf. Marchak 1983). Until the current round of economic globalization, Canadian resource industries relied on volumes to influence pricing. This influence becomes tenuous in an accelerating global economy, and Canada's resource economies are increasingly being undercut in the low-value market they once dominated – trapped between high labour costs, environmental limitations, and a paucity of secondary or value-added manufacturing capacities (Burda and Gale 1998; Howlett and Brownsey 2008).

Thus, economic globalization is accelerating challenges and trends that have been evident for some time, but in a way that is leading to significant disjunctures in rural and resource economies in Canada. To date, this has been most evident in forestry and fisheries, and we will briefly consider recent changes to these sectors because of important parallels to developments in Canadian aquaculture.

Forestry arguably represents the archetypal staples industry, as forest products are essential for industrial development of core regions (everything from housing to newsprint) while forest extraction activities often come to dominate peripheral communities located near the resource (Marchak 1983). This also means that forestry has been shaped by the demands of faraway markets. In particular, Canadian forestry has long depended on export to the United States (notwithstanding cyclical trade disputes), and this dependency has only increased following the Canada-US Free Trade Agreement and the North American Free Trade Agreement (NAFTA). For instance, in 1962, 74 percent of Canadian exports of forest products went to the United States; in 2005, the figure stood at 81 percent (Hayter 2000; Statistics Canada 2006). While exports to Europe and Japan have fluctuated over time (exports to Japan reached as high as 19 percent of total exports in 1987), in 2005 they constituted 6 percent and 4 percent of total exports, respectively.

Paradoxically, then, for Canadian forestry globalization has meant a closer attachment to the United States. At the same time, however, trade liberalization has encouraged new *producing* regions to enter the global marketplace and effectively compete with Canadian industry. Burda and Gale (1998) point to South America as the primary threat to Canadian dominance of the US market, particularly as transport and tariff barriers are removed. Canada's position is also being weakened by changes to forest management practices in other parts of the world. Specifically, "plantation forestry" is becoming increasingly common in warmer climates, including southeastern regions of the United States (Sedjo 1999; Smith et al. 2001). Plantation forests are intensively managed and highly productive tree farms that are modelled on agricultural principles; they now account for nearly one-third of global forest harvest (Bael and Sedjo 2006).

The entry of new producers has driven forestry exports to new heights worldwide, nearly doubling from 1985 to 2005 (FAO 2007a). At the same time, however, Canada's share of world forestry exports has fallen from 23 percent in 1984 to 16 percent in 2005. This presents Canadian producers with an escalating problem: as producers lose market share, they further lose the ability to influence pricing, a situation that places Canada's traditional high-volume, low-cost forest sector in a very precarious and volatile position (Burda and Gale 1998). Economic globalization is not radically changing the industry in Canada: this country still overwhelmingly exports minimally processed, standardized forest products. Globalization accelerates competition, however, and paradoxically makes Canada increasingly dependent on the US because that is where Canadian producers are best able to compete in an increasingly difficult global market (Hayter 2000, 233).

The globalization of world fisheries has also caused a relative decline in Canada's world presence. In the 1950s, Canada generated 5 percent of the world's seafood production. By 2000, however, this had fallen to less than 1 percent (FAO 2007b). This relative decline masks real growth: from 1950 to 2004, Canada's fish production increased 140 percent from all sources (fisheries and aquaculture), while world fish production rose an astounding

785 percent. Nevertheless, Canada's declining share of world fish production means that Canadian producers are again losing competitive advantage and the ability to influence pricing. Somewhat ironically, this loss of influence is one of the reasons often given by government officials to justify aquaculture expansion (e.g., Bastien 2004, 8), particularly since Canada's two most lucrative fisheries – Atlantic cod and Pacific salmon – have become shadows of their former selves. The well-documented collapse of the North Atlantic cod fishery has cost Canada its dominance of an important world fish market, and Canadian grocers now import cod from countries such as Norway, Russia, and Iceland (OECD 2003, 47). On the Pacific coast, the salmon fishery has been severely affected by conservation efforts. In the mid-1980s, Pacific fisheries typically yielded 80,000 to 100,000 tonnes annually (yearly totals would fluctuate because salmon populations are cyclical). From 1998 to 2006, however, annual catch has averaged only 26,500 tonnes (DFO 2007). At the same time, salmon fishers have seen a significant decline in prices that is directly attributable to the worldwide growth of salmon aquaculture (as we will discuss later). Overall, Canadian exports of wild-capture salmon declined 57 percent from 1988 to 2001 (BC Salmon Marketing Council 2001).

The Canadian aquaculture industry faces many of the same challenges as forestry and fisheries. Overall, world markets for commodities and food products have become increasingly complex and competitive (Bonnen 2000). This global reality has had a formative impact on industrial aquaculture. While most of the world's aquaculture products are marketed locally (particularly freshwater species), "luxury" agri-food products such as aquacultured salmon and shellfish are global commodities. Since the late 1980s, these markets have exhibited significant volatility, with world salmon production in particular growing exponentially while average prices dropped from US$6.10 per kilogram in 1988 to US$3.20 per kilogram in 2004 (without discounting inflation; FAO 2004, 50).[2] These declines in prices can be clearly attributed to the expansion of salmon aquaculture in traditional regions such as Norway and the United Kingdom, as well as the entry of new producers such as Canada and Chile. There are distinct parallels with forestry in this regard (see above), as the entry of new producing countries has caused the export value of aquacultured salmon from Norway (the traditional world leader) to fall nearly 75 percent from 1985 to 2004 (Knapp et al. 2007, 69).

The impact of this price drop has been unevenly distributed. Indisputably, the primary victims have been commercial fishers. In traditional salmon markets, poor harvest years have often meant higher salmon prices, a situation that partially compensated for this misfortune. The current dominance of aquaculture in world salmon markets means, however, that commercial fisheries have less influence over pricing, making them more vulnerable to price fluctuation (cf. Knapp et al. 2007, 215). This has led to significant conflicts between commercial fishers and the aquaculture industry in Canada, particularly in British Columbia. The price drop in world salmon markets has also affected aquaculture producing regions differently. For instance, operations in Norway have been able to reduce their costs of production sufficiently to stabilize and even increase profit margins despite the price decline, largely through a very strong research and development program (Knapp et al. 2007, 69). By comparison, Chile has become the second-largest producer of farmed salmon (behind Norway) on the strength of low costs for labour, feed, and environmental compliance. Such advantages have permitted Chilean producers to increase salmon production by an average of 42 percent annually since 1984 (Knapp et al. 2007, 66).

Canadian aquaculture producers thus face a difficult global market. Despite significant growth in the Canadian aquaculture sector (at an average annual rate of roughly 9 percent since the early 1970s), Canadian producers have trouble competing with low-cost producers. For instance, in 2005, Chilean producers were able to offer prices for fresh and frozen salmon fillet products at an average of US$2.20 per kilogram lower than Canadian producers (CAIA 2005a). This market reality has in part motivated investments by federal and provincial governments in aquaculture research and development in an attempt to make Canada a leader in aquaculture innovation and thus take the high road of value-added research and production into global markets rather than the low road of reliance on cost advantage (see Chapter 7). Arguably, the biggest threat to the Canadian aquaculture industry is that it will be caught between the two models represented by Norway and Chile.

The world market for aquacultured salmon is dominated by a small number of multinational firms that locate production in multiple regions of the world (Phyne and Mansilla 2003). At the turn of the millennium, world salmon production was dominated by a half-dozen multinational companies. A recent round of mergers and acquisitions has reduced this number to two firms headquartered in Norway: Pan Fish (which acquired rival giants Marine Harvest and Fjord Seafoods in 2006), and Cermaq (which acquired George Weston Ltd.'s Pacific Canadian aquaculture operations in 2005). Both firms now have extensive holdings across the major salmon-producing regions of Norway, the United Kingdom, Canada, and Chile. This means that major salmon aquaculture firms are able to draw value out of both the high and low roads to world markets. The role of Canada as a producing region is unclear in this regard. While Canadian shorelines are well suited for aquaculture production, resistance to the industry has hindered growth. At the same time, Chile has been successful in parlaying low-cost advantages into dramatic increases in production as well as a dominant position in value-added processing. Low labour costs in particular have enabled major firms to expand processing operations in Chile, to the point that Chilean production dominates the import market for fresh and frozen salmon fillets in the United States, and is growing exponentially in market share for canned salmon (Knapp et al. 2007, 141). Canadian producers are generally unable to compete in these sectors, and are concentrating primarily on the minimally processed fresh market in Canada and the United States (CAIA 2005a).

Caught between high and low roads, the Canadian aquaculture industry has followed the path of so many prior resource sectors and become strongly dependent on export to the United States. As of 2005, Canada was the fourth-largest producer of aquacultured salmon (behind Norway, Chile, and the United Kingdom), and the United States received 93 percent of Canadian exports (CAIA 2005a). Clearly, the American import market, valued at US$1.2 billion, is crucial for the survival of the Canadian salmon aquaculture industry. Aquaculture differs in a key respect from traditional resource sectors such as forestry, however, in that Canada is a relatively small player in the American salmon market. Chile is by far the largest player in US imports, holding a 60 percent share of the US market in 2004 (CAIA 2005a). In 2006, Chilean producers moved an estimated 72,000 tonnes of fresh salmon to the United States, compared with an estimated 7,000 tonnes from Canadian producers (*Fish Farmer* 2006). Moreover, recent fluctuations in currency have cost Canadian aquaculturists one of their key competitive advantages. Throughout the 1990s and into the new millennium, Canadian producers benefited from a weak Canadian currency (valued at between 60 and 70 cents to one US dollar). This magnified profits while enabling Canadian firms to remain marginally cost-competitive in the United States (PWC 2003). By late 2007, however, the Canadian dollar had climbed to par with the US dollar, thus increasing cost/price pressures on Canadian producers.

This strong but precarious dependency on a single export market raises legitimate fears that Canada's aquaculture sector is replicating the "staples trap" that has characterized traditional Canadian resource development, where the industry focuses on a single product and market and fails to invest in the diversification necessary for long-term resiliency (Watkins 1982). These questions are compounded by the fact that Canadian producers have only a minor presence in the US salmon market, and therefore have very little influence over pricing. So although there are important parallels with traditional sectors such as forestry, a crucial difference is that the Canadian aquaculture industry has emerged in an era of more complete globalization, where Third World producers are able to leverage low-cost advantages into significant market share. Perhaps the clearest indication of Canada's small stature in US salmon import markets is the fact that Chilean producers have been the target of protectionist lobbies in the United States since 1997. In a move that eerily parallels the Canada-US softwood lumber dispute (a US import market in which Canada is the major player), a group called the US Coalition for Fair Atlantic Salmon Trade filed a petition with the Department of Commerce in 1997, accusing Chilean producers of dumping low-cost salmon on the US market as well as fixing prices below the costs of production (Knapp et al. 2007, 244).

In summary, globalization both defines and challenges the Canadian aquaculture industry. Some of these challenges parallel those facing Canada's traditional resource sectors, and some stem directly from the late emergence of the industry in a context of intense global competition (meaning that it has never been a major player in world export markets). Canada's strong reliance on minimally processed exports to a single market suggests that Canada's aquaculture sector is significantly more vulnerable to market and currency swings than other producing regions in either the First or Third Worlds, which are better able to influence pricing and are more invested in value-added activities. Indeed, in recent years Canada's aquaculture industry has shown troubling signs of year-to-year boom-and-bust swings based largely on short-term trends in pricing and consumer demand (CAIA 2005a; Harvey 2006, 11). Globalization means that the industry has exhibited both significant growth and deep vulnerabilities, and its economic future is uncertain. The Canadian aquaculture industry has weathered intense controversy over the past decade and a half, but questions about its long-term economic viability remain, particularly as competing regions continue to expand aggressively. Thus, the stakes are high in the ongoing political debates over the fate of the aquaculture industry in Canada.

Neoliberal Governance and Development
Just as the Canadian aquaculture industry (and controversy) is being shaped in the crucible of economic globalization, it is also entangled in significant changes to how rural and resource economies are governed and regulated. Broadly speaking, these changes are part of a global (but highly varied) movement in political philosophy and policy away from what is often termed "Fordist-Keynesian" strategies for economic development and regulation, and towards neoliberal strategies for achieving these (we will explain these terms in a moment). Aquaculture is the first major resource industry in Canada to grow up in the context of neoliberal ideologies and policy regimes. As we shall see, other major sectors, such as forestry, were formed under very different economic and environmental regimes, and are also currently undergoing controversial transitions away from Fordist-Keynesian practices and towards neoliberal forms.

Neoliberalism is the major political economic orthodoxy in the world today (Peck and Tickell 1995, 2002). It is both a worldview and a specific way of organizing governance and policy (Jessop 2002a). Neoliberalism is simultaneously a universalist doctrine, with academic and political adherents who believe that it can be applied nearly anywhere to any set of policy problems, and a flexible doctrine that is applied very differently around the world (Larner 2003). Above all, neoliberalism is the political and regulatory foundation of economic globalization. Like the term "globalization," however, "neoliberalism" is a slippery concept that is sometimes used carelessly or haphazardly. Some academic critics have questioned the utility of the term, particularly if it is used only as a shorthand for any and all government and corporate actions that are not identifiable as progressive or leftist (see Barnett 2005; Castree 2006). On the other hand, some scholars argue that the concept of neoliberalism offers one of the most useful tools for understanding contemporary governance – going even so far as to say that "neoliberalism appears to have usurped globalization as *the* explanatory term for contemporary forms of economic restructuring" (Larner 2003, 509). Adherents of the latter position have worked to build a theoretical basis for understanding neoliberalism both as a precise set of ideological principles and as a worldwide political movement that is flexible and adaptive to different circumstances – equally but differently applicable to the agricultural economies of Latin America and the urban cores of Europe and North America (e.g., Perreault and Martin 2005; Brenner and Theodore 2005).

Although formal definitions of neoliberalism vary, in our view the conceptualization of neoliberalism from the perspective of regulation theory is the most promising for understanding both the precision and flexibility of this movement (Young and Matthews 2007b). Regulation theory draws on the iconic thinking of Karl Marx, Antonio Gramsci, and Nicos Poulantzas, each of whom sought to understand the relationships between capitalist economies, governments, and societies. For each thinker, one of the most profound puzzles of the modern world is the endurance of capitalist economies in the face their own contradictions (Jessop 2002a, 6). These contradictions are economic (in the tendency of capitalist markets to boom and bust, to concentrate capital and wealth, and to institutionalize inequality and unemployment) and political (in that the real winners in capitalist economies are few in numbers but very politically powerful). The puzzle, then, involves inquiring into the source of capitalism's resiliency; in other words, how has capitalism survived its economic and political tendencies, and how has an inherently volatile and unequal system won the consent of those who are (ostensibly) its victims? These are big questions, and Marx, Gramsci, and Poulantzas each wrote voluminously on the subject (e.g., Marx 1985 [1848]; Gramsci 1992 [1935]; Poulantzas 1971). Again, while specific interpretations vary, the core point that has been carried from these classic authors into contemporary regulation theory is that capitalism operates through the marriage of *methods of accumulation* and *methods of social regulation*. In other words, regulation theory proposes that one of the key roles of government in a capitalist economy is to negotiate the relationship between economic and social policies in the interests of stabilizing the predominant macroeconomic system of the day.

To illustrate, we draw on the work of Jessop (2002a, 2002b, 2004), a leading figure in the regulationist school. Jessop argues that there are identifiable eras in the history of capitalism where governments have sought different "fixes" for the contradictions of capitalist markets and practices. Simply put, as capitalism evolves, so do the ways in which governments regulate and support it. In this light, any argument that capitalism is changing (an argument that is front and centre in the current literature on globalization) prompts us to look closely at state activities. Indeed, Jessop's main concern is with understanding the current period as one of transition between two eras in the history of world capitalism. According to Jessop, the mode of regulation that is currently receding is that of the Fordist-Keynesian state, or the "model of the postwar state in the United States, Canada, Northwestern Europe, Australia, and New Zealand" (Jessop 2002a, 55).[3]

Taking a step back, Fordism is essentially a method of production that is based on the mass production of standardized products (the term being directly derived from the Ford Motor Company's pioneering of assembly line production) that are typically destined for national or continental markets. The imprint of Fordism lies strongly on the Canadian economy, it being the anchor for postwar development of the country's manufacturing economy (particularly in the automobile sector and in textiles) as well as resource economies (cf. Hayter and Barnes 1997; Wallace 2002). Keynesianism generally refers to the dominant method of government participation in national and regional economies during the Fordist era. Named for the British economist John Maynard Keynes, Keynesianism refers to the idea that governments ought to play a strong role in stabilizing the economy by supporting consumption activities (Johnson 1971). Thus, a Fordist-Keynesian economy

> can briefly be defined as an accumulation regime based on a virtuous auto-centric circle of mass production and mass consumption secured through a distinctive mode of regulation expressed in a [Keynesian] state ... In this form, the state contributes to the delicate balance of production and consumption by helping to integrate the circuits of the capital and consumer goods industries by managing the conflicts between capital and labour over both the individual and the social wage so that the virtuous circle of Fordist growth can be maintained. (Jessop 2002a, 55, 57)

Figure 1.1 provides a general summary of the strategies employed by states in the Fordist-Keynesian era in their attempts to stabilize Fordist capitalism. These attempts involved significant government activism, particularly in ensuring growth in both supply and demand. This arrangement anchored the "long boom" experienced by Fordist-Keynesian nations from the postwar period until the mid to late 1970s. The crisis in Fordist-Keynesian economies that emerged in the 1970s and 1980s has been the subject of lengthy academic debate (e.g., Lipietz 1987; Lash and Urry 1987; Teeple 1995; Brenner 2004). While opinions vary, this literature argues that the marriage of mass production and consumption *within* national economies became increasingly untenable with time and growth (leading to inflation and recession), and increasingly fragile in the face of shocks such as the 1973-74 oil crisis (cf. Brenner 2004).

The Fordist-Keynesian economy and state did not collapse quickly, nor have they receded completely. Nevertheless, the crisis in Fordist-Keynesian arrangements has prompted an extensive and broadly based set of economic reforms across both advanced capitalist and developing countries towards neoliberal policies for economic and social governance. These reforms were initially led by true believers in state retrenchment and deregulation, such as Prime Minister Margaret Thatcher, President Ronald Reagan, and (to a lesser extent) Prime Minister Brian Mulroney, during the late 1970s and 1980s, but in recent years the movement to neoliberalism has been joined by governments of centrist and even leftist leanings. Indeed, Brenner and Theodore (2002) argue that every major government in the advanced capitalist world has now embraced neoliberalism as a guiding economic philosophy.

Neoliberalism is a highly varied movement, but its fundamental principles can be summarized as shown also in Figure 1.1. To understand this summary, it is important to recall that neoliberalism is a political response to the transition from a regime of accumulation based largely on Fordist production within national boundaries to a regime that is linked closely with economic globalization and its imperatives for internationally organized production and consumption. If the aim of state intervention under Fordist-Keynesianism was to stabilize and balance production and consumption within nations, the aim of neoliberal state strategies is to promote flexibility and competitiveness in a wider economic environment (Jessop 2002a; Brenner 2004). Thus, as outlined in Figure 1.1, a significant component of neoliberal policy is to encourage innovation and extend corporate rights while permitting downward pressure on wages (in some sectors and occupations) and the subordination of social policy to economic policy.

In addition, neoliberal governance encompasses a very different approach to economic development than in Fordist-Keynesian regimes. Brenner (2004, 134) argues that in the Fordist-Keynesian era, economic development was equated with the drive "to maximize national output and income by promoting a balanced spatial distribution of socio-economic capacities and infrastructural investments across the entire national territory." In other words, if Fordist-Keynesian economies were based on the mutual development of production and consumption capacities, then "significant large-scale territorial disparities were viewed as a major threat to stabilized patterns

Figure 1.1

Fordist-Keynesian and neoliberal states

Characteristic	Fordist-Keynesian state	Neoliberal state
Distinctive set of economic policies	Full employment; demand management; provision of infrastructure to support mass production and consumption	Focuses on innovation and competitiveness in open economies
Distinctive set of social policies	Collective bargaining; generalization of mass consumption; expansion of welfare rights	Subordinates social policy to an expanded notion of economic policy; downward pressure on the "social wage"
Primary space or scale	Relative primacy of national scale in economic and social policy making, with local as well as central delivery of service	Multiplication of scales above and below the national scale, but national states continue to have a strong regulatory role
Primary means to compensate for market failure	Market and state form a "mixed" economy; the state is expected to compensate for market failures	Increased role of self-organizing governance to correct for both market and state failures

Source: Adapted from Jessop 2002a, 59, 252.

of macroeconomic growth" (Brenner 2004, 130). Territorial inequalities were indeed a substantial preoccupation of Canadian governments during the postwar period. At the federal level, this concern led to the creation of massive regional development programs in the 1960s and 1970s, such as the Department of Regional Economic Expansion (cf. Savoie 1992). Many provinces also invested in this ideal in their attempts to spread Fordist-Keynesian institutions across rural and remote regions (we will consider this below).

With economic globalization, the need to promote geographically even or universal development in order to "maintain the virtuous circle of Fordist growth" recedes (Jessop 2002a, 57). Brenner (2003) argues that globalization shifts the axis of economic development policy away from horizontal concerns (geographic uniformity or equity) and towards vertical or scalar concerns (the links between particular localities and global markets and flows). Thus, neoliberal economic development policies explicitly aim to foster *direct linkages* between local and global as the engine for national economic growth: "[Against] the project of equalizing the distribution of industry, population and infrastructure across national territories, [neoliberal] states strive to differentiate national political-economic space through a reconcentration of economic capacities into strategic urban and regional growth centres" (Brenner 2003, 206).

In the Fordist-Keynesian era, governments asserted their vision of economic development by centralizing authority over industrial development, expanding the institutions of the welfare state, and playing an active role in addressing market failure and gaps through institutions such as Crown corporations (see Figure 1.1). Neoliberal governance uses a very different set of tools for promoting "glocal" or geographically variable development. In general, these tools involve the *selective devolution* or partial transfer of authority and/or responsibility to private actors (corporations, groups, or individuals) (Jessop 2002b, 454). Devolution is a complex policy tool. In some cases, it can be described using the familiar concepts of privatization, deregulation, and liberalization. At other times, devolution is more controlled or highly selective, where governments transfer authority in a context that encourages private actors to "act rightly" or to exercise their new freedoms in a manner that reflects the goals of "post-welfare" governance (such as self-sufficiency, self-discipline, and efficiency; cf. Herbert-Cheshire and Higgins 2004).

This shift in development strategies has been well studied in urban settings (Harvey 1989; Graham and Marvin 2001; Jessop 2002b; Brenner and Theodore 2005), but it also has strong implications for rural and resource regions. First, it has meant that federal and provincial governments in Canada have much less interest in making investments in rural regions for the sake of addressing regional inequalities. Across the country, rural services are being reduced, particularly social and welfare services, as well as centres of administration such as management offices for forestry, agriculture, and parks (Epp and Whitson 2001; Halseth and Halseth 2004; Young 2008). Second, the shift to neoliberal development strategies in Canada has meant a significant devolution of authority to resource firms with respect to discretion over economic geographies, environment, and labour relations (Young and Matthews 2007b; Young 2008). In the Fordist-Keynesian era, federal and provincial governments pursued rural and resource development agendas based on efforts to foster large-scale production that was as geographically dispersed as possible. This "geography of Fordism" was built through direct government investment in infrastructure and community building, as well as the imposition of conditions on rights to publicly held resources that required major resource firms to process core commodities (particularly wood and fish) in designated facilities and/or regions with the aim of stabilizing rural economies (cf. Hayter 2000; Young 2008). These strategies were often explicitly mobilized with the deeply Fordist-Keynesian intent of province building, where vibrant resource exploitation dispersed across the territories and communities of the province was understood as a key anchor in the "virtuous circle" of production and consumption capacities and therefore of overall prosperity and development (cf. Barman 1996; Sandberg and Clancy 2000).

With economic globalization and the crisis in Fordist-Keynesian economies, governments have moved to permit greater flexibility in resource production. For example, in British Columbia's major resource sectors, this has included policy reforms such as the elimination of spatial restrictions on processing (i.e., the conditions that linked processing with the region of harvest) as well as more flexible environmental regulation (Young and Matthews 2007b). Similar reforms are being considered in Quebec and Ontario, both of which have suffered severe downturns in their forest sectors (Dutrisac 2008). Reforms such as these represent a final rejection of the Fordist aim of reducing territorial disparities as the primary objective of development policy. This is made clear in the following statement from the government of British Columbia regarding the logic behind the elimination of spatial restrictions on processing in forestry:

> Timber processing [restrictions] were introduced in an attempt to create local or regional economic benefits from the timber that was [locally] logged. But these regulations led to a series of unintended consequences that hinder the forest sector's ability to make sound, business-based decisions ... Forcing licensees to process wood at mills with equipment that is outdated, or at mills that make products that are not in demand, prevents valuable public timber from flowing to other, better uses ... Some British Columbians view these policies as part of the social contract that forest companies should meet in exchange for the right to log public land. But while these policies may have made sense in a different time with different market conditions, they have not shielded today's communities from job loss and economic difficulties. In fact, they serve as a disincentive and impediment for the forest industry. (British Columbia 2003a, 17)

In ending geographical restrictions on processing, governments are essentially granting licence for investment and jobs to pool in regions and communities that are best positioned to compete at a global level. Thus, as in urban areas, the ideal of universal development is giving way to one where the concentration of investment in particular localities is considered essential and desirable. What this means, however, is that areas that are not seen to be ideally positioned for global competition in resource sectors are increasingly vulnerable to further decline. Indeed, current research suggests that *coastal regions* are being particularly impacted by these changes in policy as the geography of fish and forestry processing shifts away from smaller communities (Young 2008).

Neoliberal strategies for economic development also envision a new role for local communities. In the Fordist-Keynesian era, control over local development was in fact highly centralized in the hands of senior governments and major employers (Matthews 1983). Local governments were not typically involved in questions of economic development, and generally focused on service delivery and administration (Hayter 2000, 288). Neoliberal strategies break from this tradition. First, as we have discussed, devolution to corporate actors is freeing them from traditional obligations to resource communities, thus destabilizing the traditional system for local development. Second, in recent years both federal and provincial governments have created new rural development programs that encourage entrepreneurialism among local governments, groups, and firms. These programs – which include the Community Futures network (Community Futures Development Corporations in Western Canada and Ontario, les sociétés d'aide au collectivités au Québec, and Community Business Development Corporations in Atlantic Canada), various programs of Western Economic Diversification (WED) and the Atlantic Opportunities Agency (AOA), as well as various provincial programs such as Community Forestry – provide a means for local actors to pursue from-the-ground-up initiatives for local development. Importantly, this strategy for local development reflects *both* social democratic and neoliberal ideologies (McCarthy 2005; Young 2008). On the one hand, the move to support ground-up development is a significant step towards community empowerment and local control over directions for development. On the other hand, this strategy meets the neoliberal goal of replacing the paternalism of universal programs with locally driven initiatives that are often led by champion groups and evaluated through competitive review, making the expenditure of public funds for development much more targeted and efficient (Graham and Marvin 2001, 309). Moreover, this development strategy encourages localities to pursue particular advantages that, as Brenner (2003, 2004) argues, are thought in the neoliberal discourse to be the cornerstone for success in the global economy.

Thus, in rural and resource economies, the neoliberal approach to economic development involves two main movements: the (further) empowerment of major corporate actors and the (modest) empowerment of local actors to pursue local development goals (Young 2007).[4]

We argue that state approaches to aquaculture in Canada involve both types of devolution. Federal and provincial governments in Canada have sought to selectively devolve key aspects of industrial development to aquaculture firms and to communities. To be clear, aquaculture remains a highly regulated industry, particularly with respect to salmon and other finfish.

The nature of these regulations and their enforcement appears to be moving, however, in the direction of what is termed "results-based regulation" or "performance-based regulation" (Howlett and Rayner 2004). Proponents of results-based regulation argue that government regulations ought to focus on establishing standards and thresholds for end-of-pipe outcomes rather than engaging in "prescriptive" or "process-based" regulations that seek to govern the private sector by monitoring its activities step by step (cf. Coglianese 2003). This approach is based on two assumptions. The first assumption is that the private sector is inherently innovative, and if unshackled from prescriptive regulations, it will find efficiencies along the way to meeting good, scientifically determined end results (Vold 2003). The second assumption is that firms operating in a global economy require the flexibility to bend the rules at times, so long as things are set right in the final assessment (West Coast Environmental Law 2002). Of all jurisdictions, British Columbia has been the most aggressive in drafting and implementing results-based regulations for aquaculture (British Columbia 2006b). We will return to this question in Chapter 7, but for the present discussion it is important to note that this shift in regulatory approach grants significant authority and flexibility to corporate actors to manipulate the details of production and environment, and that state authorities deem this flexibility to be crucial to the ability of Canadian aquaculture operations to compete with low-cost producers in a global market (Canada 2006).

Governments are also moving to involve communities in small-scale aquaculture at the same time that they strengthen corporate power and authority over large-scale operations. For instance, in a distinct parallel to community forestry programs, the government of British Columbia has developed a program called the Shellfish Development Initiative, which "works with coastal and First Nations communities" in providing knowledge and logistical support for locally driven shellfish aquaculture (British Columbia 2007a). The federal government has also been active in supporting the development of local and community-driven shellfish aquaculture (cf. Canada 2001; Rayner and Howlett 2008). The promotion of shellfish aquaculture at the community level makes economic sense, as it is significantly less capital-intensive than finfish aquaculture. According to Tollefson and Scott (2006), however, government promotion of shellfish aquaculture at the local level is politically strategic, as the practices and principles of community-based shellfish aquaculture are in some ways consistent with traditional Aboriginal usage of the resource (particularly in terms of caring for shellfish beds, as well as the principles of group ownership and custody).

We do not suggest that governments in Canada are promoting shellfish aquaculture at the local level solely as a means of justifying the expansion of corporate-dominated finfish aquaculture. There is little if any evidence of such a strategy. From the perspective of regulation theory, however, the coincidence of devolution to community actors and to corporate actors is important, particularly because the latter is occurring on a much grander scale and over higher stakes (Young and Matthews 2007b; Young 2007). Devolution to corporate actors (in sectors with revenues in the hundreds of millions and even billions) characterizes the predominant *method of accumulation* in neoliberal resource regions. This prompts us to ask whether devolution to communities and increased state support for endogenous development are the other side of the neoliberal coin (a form of *social regulation* or compromise). In the Fordist-Keynesian era, the chief compromise for macroeconomic stability in resource regions involved collective bargaining and the extension of welfare-state institutions to the periphery (cf. Marchak 1983). In the emerging era of economic globalization and neoliberal governance, the compromise now appears to be based on the pairing of a liberalized corporate resource economy with a commitment to support ground-up, entrepreneurial, community-level development.[5]

The shift to neoliberal governance and strategies for economic development complicate the aquaculture controversy. As we will discuss in Chapter 2, many coastal community members and Aboriginal groups reject the implied neoliberal compromise of granting greater discretion to resource firms. While community empowerment is generally welcomed, many are deeply skeptical of the further extension of corporate authority over resource spaces. Moreover, as we will see in Chapters 6 and 7, state efforts to foster regional advantages and local/global linkages may be intensifying the conflict over aquaculture development. For instance, state programs such as WED and AOA, as well as arm's-length organizations such as Community Futures, have strongly endorsed tourism development in coastal areas. As industries such as tourism expand, so do potential conflicts with aquaculture, particularly over pollution, threats to the lucrative sport fishery, visual aesthetics, and the enclosure of navigable waters. The Fordist-Keynesian compromise came with real costs, particularly in the form of systematic environmental degradation and in the dependency of rural labour and communities on centralized authorities (Marchak 1983; Matthews 1983). In some ways, the emerging neoliberal compromise for resource economies and regions addresses these problems, but also creates new tensions in fostering flexible, diversified, "glocal" rural economies.

The Global Environmentalist Movement
The third global force that is profoundly shaping the aquaculture industry and controversy in Canada is the global environmentalist movement. This movement is based on a worldview that stresses the tight coupling of local and global ecologies, rights, and political issues. This paradigm is held and expressed across multiple activist movements, some of which are globally

organized while others are locally focused but deeply embedded in the globalist discourse. To begin, it is important to recognize that although the notion of a "global environment" is familiar to us now, this is a relatively new conceptualization, at least in its current place as a major cultural and political touchstone. As we will discuss below, some scholars suggest that the true birth of an identifiable "global environmental movement" occurred as late as the 1980s and 1990s, when issues such as nuclear safety, atmospheric ozone depletion, and global climate change began to dominate thinking about environmental health and integrity. If this is the case, it means once again that aquaculture has emerged under a very different set of political and cultural circumstances than Canada's other major primary sectors. In this section, we present a brief overview of the emergence and shape of the global environmentalist movement, as well as some of its consequences for aquaculture development in Canada.

The rise of the contemporary environmentalist movement has been nothing short of remarkable. In a span of only a few decades, it has moved from the fringes to the centre of nearly every debate on industrial development. In this short time, environmentalist groups have successfully challenged governments and corporate actors on local, national, and global scales, often using controversial and confrontational means, while at the same time gaining and building public trust as authoritative voices on complex scientific issues (Yearley 2003, 41).

While the contemporary environmentalist movement has grown exponentially in recent years, it has deep roots and a complex history. Several scholars have noted the debt owed by modern environmentalism to religious mysticism as well as to the anti-modernist romanticism that emerged following the Enlightenment of the seventeenth and eighteenth centuries (cf. Cronon 1995; Palmer and Cooper 1998). Environmentalism in Europe and North America has a direct genealogical link to the preservationist movement that emerged in the nineteenth century. According to Macnaghten and Urry (1998, 36), "preservationism was a Victorian reaction against the Enlightenment mentality which assumed that nature was to be improved through human reason and interference." In particular, the preservationist movement was a reaction against the spread of industry and urban development, arguing that built environments were unnatural and artificial spaces that impugned God's cathedral in the form of natural and wild landscapes (Cronon 1995). Preservationism resonated with many members of the nineteenth-century intelligentsia, and was instrumental in establishing national parks and other protected wilderness areas, particularly in the United States and Canada. The preservationist movement continues to provide some of the most vivid cultural imagery regarding human encounters with wilderness and raw nature, particularly as represented in the writings of Henry David Thoreau and John Muir.

Despite these enduring legacies, preservationism was ultimately a limited social movement that succumbed to a more pragmatic "conservationism" in the early twentieth century (cf. Mazzotta and Kline 1995). While conservationism is itself a varied movement, generally speaking it draws on a rational and scientific view of the environment, specifically "a utilitarian desire to regulate nature through rational and efficient management" (Macnaghten and Urry 1998, 34). Conservationism has an ambiguous legacy in current environmentalism. On the one hand, one of the principal limitations of earlier preservationism was that it was imagined to apply chiefly in "special" places – areas that were to be set apart and protected from the march of human progress. In this sense, conservationism significantly broadened the scope of environmental management, being based on the understanding that natural resources were not inexhaustible and required regulation and protection from reckless development and exploitation. Indeed, the dominant postwar resource management strategy in North America has been conservationist, based as it is on the principle that limitations must be placed on natural exploitation, a view that was radical in its day (Hayter 2000). On the other hand, conservationism has led down some very destructive paths. The idea of rational management implies a certain hubris with respect to human capacities to know and predict the natural environment. We need only look to the Canadian experience with the collapse of the Atlantic cod fishery to appreciate the costs of incomplete knowledge and an overly pragmatic approach to environmental management.

The current environmentalist movement continues to grapple with the issues raised by the preservationist and conservationist perspectives, as evidenced by ongoing debates regarding the appropriateness of "sustainability" as an environmental narrative (given its close association with rational management). What distinguishes the present environmentalist movement from prior perspectives and activism is its focus on global issues, and more precisely the intersection of local and global environmental challenges. As mentioned, preservationism was primarily directed at specific spaces. In Macnaghten and Urry's words (1998, 37), it was a movement "to regulate boundaries, especially between town and country" – the realm of humanity and that of wilderness (see also Cronon 1995). While conservationism broadened the scope of environmental discourse and management, the emphasis on boundaries remained, given that natural environments were to be managed in the best interests of regions, states and provinces, and nations. According to McCormick (1995, 56), the emergence of a "New Environmentalism" implied a change both in the scale of environmental consciousness and in the core problematic posed by the human/nature dynamic: "If nature preservation had been a moral crusade centred on the non-human environment, and conservation a utilitarian movement based on the rational management of natural resources, New Environmentalism addressed the entire human environment. For preservationists, the issue was wildlife and habitats; for conservationists, the issue was natural resources; for the New Environmentalists, human survival itself was at stake."

McCormick's concept of a New Environmentalism addresses two key shifts in the environmentalist movement that began in the 1960s: the emergence of a global consciousness and a heightened concern with dramatic or apocalyptic environmental change. Concern with global problems and concern with apocalyptic scenarios went (and still go) hand in hand. Macnaghten and Urry (1998) point to the publication of Rachel Carson's *Silent Spring* in 1962 as a particularly important watershed moment in modern environmentalism. Carson's book, which sold 500,000 copies, outlined the consequences of industrial usage of pesticides on wildlife, particularly among animals occupying high places on the food chain, which would be exposed to damaging and even lethal concentrations of chemicals. For Carson, the boundaries of place, region, nation, and continent that had been presumed under preservationist and conservationist conceptions of environment were nonsensical, given the cumulative dangers unleashed by unfettered global industrial development. In short, the globality of the problem spoke directly to its apocalyptic potential. This was not a problem that could be fixed or regulated in any one place. Even the metaphor of a "silent spring" implied the mobility and transnationality of this problem, as toxins migrated along with their avian victims. According to Macnaghten and Urry (1998, 45):

> Carson painted a picture of a world in mortal danger, a danger systematically and cynically produced by the greed and self-interest of the pesticides industry. Even more significant was the diagnosis that these "elixirs of death" which entered the human body were a direct by-product of the post-war zeal for modernization and technological improvement. While previous concerns had centred on the aesthetics of suburbanization, or local pollution incidents, or the loss of particular habitats, Carson's critique centred on a representation of nature as systematically threatened by modern industrial processes.

Carson was joined by other prophets of doom, who used strong scientific language to make their cases (Macnaghten and Urry 1998, 45). Among the most influential of these were Paul Ehrlich (on world population growth), Garrett Hardin (on "the tragedy of the commons"), and Barry Commoner (on pollution and toxins). According to Jasanoff (2004), at the same time that discussion of such global problems was entering the public and political arena, humanity's environmental imagination was forever changed by the dissemination of some of the most iconic images ever taken – those of Earth from space. These images, in which political boundaries are both imperceptible and irrelevant on a small blue and white ball, have had a profound cultural impact. In the words of astronaut William Anders, who filmed the famous "earth rise" on the Apollo 8 mission: "We came all this way to explore the moon, and the most important thing is that we discovered the Earth."

From the 1960s onward, this understanding of environmental problems as being both global and apocalyptic gained significant public and political credibility. The first Earth Day was held in April 1970 and drew hundreds of thousands of demonstrators to the streets in the United States alone (McCormick 1995, 79). The United Nations held its first major conference on environmental issues in 1972 in Stockholm, Sweden. Among the most enduring implications of this New Environmentalism is the notion that growth itself – particularly industrial development and population growth – is the primary cause of environmental degradation and risk. While preservationism and conservationism advanced criticisms of the excesses of industry, modern environmentalism takes the critique of industrial development as its core. From the beginning, this critique has been global in orientation, as is evident in the famous Club of Rome experiment in computer-modelling global environmental degradation based on contemporary trends in population growth, resource depletion, and industrial development (see Meadows et al. 1972). While the prophets of doom were criticized by contemporaries for sensationalizing their findings (e.g., Cole 1973), governments and other institutions responded in ways that reinforced the links that were being forged among the ideas of globality, growth, and the potential for apocalyptic outcomes. The most famous example of such a response was the World Commission on Environment and Development, established by the United Nations in 1983 and chaired by Gro Harlem Brundtland, former prime minister of Norway. The "Brundtland Report," entitled *Our Common Future*, embraced many of the claims of the global environmentalist movement, pointing to growth as the major threat facing the global environment and advancing the notion of sustainable development as a key to mitigating the problem. This term, here referring to "development that meets the needs of the present without compromising the ability of future generations to meet their own needs" (Brundtland 1987, 8), has since become central to nearly every debate over industrial development, including the aquaculture controversy in Canada. While the notion of sustainability has since become highly malleable, the Brundtland Report is fairly unequivocal, stating in language reflecting the clarion call of global environmentalism: "Sustainable global development requires that those who are more affluent adopt lifestyles within the planet's ecological means ... We do not pretend that the process will be easy or straightforward. Painful choices have to be made. Thus, in the final analysis, sustainable development must rest on political will" (Brundtland 1987, 9).

Since the 1980s, the global environmentalist movement has grown in size and sophistication. As previously mentioned, its greatest achievement has been its ability to link local and global issues and conflicts. This ability would not be possible without the heightened awareness of globality among the general public, experts, and institutions, as discussed above. Environmentalist groups have worked very hard to establish themselves as authorities on issues across both local and global scales, however. Environmental activism on a global scale is dominated by large, globally structured environmentalist groups such as Greenpeace, the World Wildlife Fund (WWF), and Friends of the Earth. What is singularly impressive is that these groups retain the ability to descend into very local issues – to lend their significant resources and legitimacy to issues that would otherwise be considered local, minor, or bounded conflicts.

This capability has blindsided the aquaculture industry in Canada. As mentioned in the Introduction, the aquaculture industry in Canada attracted very little attention from environmentalist groups until the mid-1980s. From then on, however, it has been subjected to highly organized and sophisticated opposition campaigns that have successfully mobilized protest and activism in both local and global spaces. In the case of aquaculture, as with many other environmentalist campaigns, the ability to act on multiple scales hinges on activist groups' strategy of forming networks and coalitions, which enhances both the reach and effectiveness of environmentalist protest (Rootes 1999). The network model enables activists to pool resources and concentrate their message in local settings. The benefits of this strategy are evident in the success of the Coastal Alliance for Aquaculture Reform (CAAR), which is composed of major British Columbia environmentalist groups that oppose aquaculture, including the David Suzuki Foundation, the Georgia Straight Alliance, and the Friends of Clayoquot Sound. While each of the member groups continues its individual campaign against the industry, the CAAR is a key vehicle for fundraising and coordination of protest efforts. As a one-issue coalition, it has been very successful in attracting media and public attention through press releases, slogans (e.g., "Farmed and Dangerous"), and street protests.

The network model also enables environmentalist groups to disseminate protest and claims well beyond local settings. For instance, organizations such as Greenpeace and the World Wildlife Fund have sponsored activities such as "dialogue conferences" and direct protests in the United States and elsewhere on the subject of Canadian aquaculture (cf. Greenpeace 2004; World Wildlife Fund 2004). Sophisticated networks and coalitions exist at the global level as well, with organizations such as the Seafood Choices Alliance (SCA) counting among its members giant environmentalist groups such Greenpeace, the WWF, the Wildlife Conservation Society, Environmental Defence, the Pew Institute, and Ecotrust among its members, as well as small and locally oriented groups such as British Columbia's Living Oceans Society and the Vancouver Aquarium. The SCA has been quite active in consumer affairs, leading boycotts of aquaculture products in the United States and pressuring restaurants and grocers to take aquacultured finfish off their menus and shelves.

Like many other industries that have found themselves suddenly pressured by extensive and aggressive opposition coalitions, Canada's aquaculture sector has a mixed record in responding to environmentalist claims and strategies (cf. Leiss 2001). As we will discuss in Chapter 3, the industry and its supporters have failed to anticipate and mitigate some key objections raised by opposition groups, particularly with respect to disease threats to wild species and the spectre of toxins in aquaculture products (Leiss and Nicol 2006). The challenge presented to Canadian aquaculture by environmentalist activism goes far beyond these specific issues, however. We argue that the ability of environmentalist groups to pursue strong agendas at both local and global scales has contributed to two profound developments in the aquaculture controversy.

First, the environmentalist movement's successes in local-scale activism has enhanced the validity of local voices. To be clear, not all local voices are environmentalist. As we will discuss in the next chapter, local perceptions and opinions on aquaculture and the environment are highly varied. Nevertheless, local-scale environmental activism has raised issues about the rights of stakeholders to say no to aquaculture even if senior governments are permitting development to proceed. In response, industry supporters are mobilizing their own pro-aquaculture grassroots movements (through groups such as Positive Aquaculture Awareness and the First Dollar Alliance). This places even more emphasis on local-level debates and conflicts.

Second, the environmentalist movement's ability to make this issue global – essentially bringing the controversy to dinner tables, grocery stores, and activist networks far removed from coastal Canada – has legitimized the participation of non-local voices in the controversy. In a sense, the globalization of the aquaculture controversy has been predictable. Several high-profile resource conflicts "went global" in the 1980s and 1990s due to the dedication of internationally connected activists, including British Columbia's "war in the woods" and the annual conflicts over seal hunting in Quebec and Newfoundland and Labrador. It is equally true, however, that very few local or regional environmental issues do in fact go global (Yearley 2005), and participants in the Canadian aquaculture industry are deeply frustrated with what many understand to be outside interference in a regional issue.

In summary, the discourse and structure of the global environmentalist movement means that the struggle over Canadian aquaculture is both local and global. The flexibility of this activism is a major challenge to the industry and its supporters, who often find themselves forced to react and rebut activist campaigns both in Canada and in key consumer markets in the United States (see Chapter 3). But the globalization of the aquaculture controversy is not just about activism. The environmentalist movement is a conduit for the circulation of claims and knowledge about aquaculture across geographic lines and political solitudes. Environmental networks bring concerns of local significance, such as the importance of wild salmon and other fisheries to Aboriginal and non-Aboriginal coastal people, into the mainstream of public consciousness. At the same time, environmentalist activism brings global, particularly urban, opposition into local spaces. For better or for worse, this greatly multiplies the voices involved in the debate over aquaculture development, as well as the range of knowledge and claims available to participants in the debate. Organized environmentalism is what has made this a truly public controversy. It has broken down traditional notions of who is a stakeholder, who is an expert, and how and where arguments are made about the future of the industry.

Aquaculture in a Global Context

The aquaculture controversy in Canada is in many ways the child of context and circumstances. Aquaculture is unlike any other resource or agricultural sector in Canada simply because of its novelty. As a new industry, it has been formed in a unique crucible, shaped by structural and cultural forces that differ significantly from the circumstances that shaped Canada's more established primary sector economies.

First, economic globalization has meant that Canadian aquaculture has started from a position of market weakness. To date, the Canadian aquaculture industry has been caught between the "high road" and "low road" to globalization. While Canadian governments have invested heavily in aquaculture research and development, the industry remains strongly dependent on the export of minimally processed consumer products to the United States (PWC 2003; CAIA 2005a). Prospects for moving up the value chain are dim, given that Third World producers such as Chile, with low costs for labour and environmental compliance, already dominate value-added markets in the US. Moreover, the recent rise of the Canadian dollar has negated one of the few cost advantages for Canadian producers. Therefore, although Canada's aquaculture sector is robust and has potential for significant growth, it is also precarious. In Rayner and Howlett's words (2008, 121), Canadian aquaculture is "caught in a vise" between the industry's need to expand to reach a critical mass necessary for survival and its heavy reliance on a limited product line in a singular export market.

Second, the rise to prominence of neoliberal approaches to economic development means that governments are less willing to pursue centralized plans and programs for development, and are increasingly decentralizing and devolving responsibility and authority for resource management and development to both corporate and local actors. This is advantageous to large aquaculture firms, as it grants them more discretion over production.

It also means, however, that communities have more at stake, and more to say, with regard to local and regional development. Local entrepreneurship, which is a central goal of neoliberal development policy, can itself become an obstacle to the aquaculture industry. As new industries such as tourism and the sport fishery mature, conflicts over environment, marine use, and aesthetics can intensify. The neoliberal vision of fostering regional advantages and "glocal" economies becomes complicated when conflicts arise about what these advantages are and what other interests and opportunities may be sacrificed in their pursuit.

Last, the emergence of global environmentalist discourses and activism has linked Canadian aquaculture to the notion of global environmental risk and problems of large-scale industrial development more generally. On the surface, Canadian aquaculture appears to be an issue of concern only in certain regions and locales. Environmental networks, however, have effectively mobilized narratives of consumer health and environmental degradation to make this a global-scale controversy. In doing so, these networks have popularized the conflict, brought significant media attention to bear on the issue, and disseminated claims about Canadian aquaculture across the nation and around the world.

2
Aquaculture in a Local Context

While Canada's aquaculture industry is being shaped by the collision of global forces, it is also strongly influenced by specific local and regional challenges. Most of the aquaculture production in Canada takes place in the Atlantic and Pacific coastal regions. These are diverse places, home to many different Aboriginal groups, resource workers, small businesses, ex-urban retirees, cottagers, and seasonal tourists. As we will see, this diversity contributes to the mixed local reaction to aquaculture. In the previous chapter, we argued that aquaculture, as a late-emerging industry, is uniquely influenced by emerging global forces. Here, we discuss how the industry is also subject to the particular local issues and tensions of rural coastal Canada.

First, we will examine how local understandings of aquaculture are framed by economic hardship and change. Since the early 1990s, Canada's Atlantic and Pacific coastal regions have been mired in an extended and severe economic downturn that is directly linked to troubles in traditional resource industries. While the aquaculture industry has sought to present itself as an economic lifeline to troubled coastal regions, we will see below that local understandings of aquaculture and its relationship to economic hardship are complex. Second, aquaculture is proceeding in a local context that is highly charged on the question of Aboriginal rights. On both the Atlantic and Pacific coasts, First Nations groups have strong but uncertain claims to rights over land and marine territories (cf. Woolford 2005; Fox 2006). While progress on these issues at the government level has been deplorably slow, the private sector is increasingly active in directly negotiating with Aboriginal groups in order to secure community support for development (Hayter 2003). As will be discussed, the evolving Aboriginal rights movement is having a significant impact on the aquaculture industry and controversy. While First Nations groups are adopting different and sometimes competing stances on aquaculture development, we argue that the issues of rights and self-governance are front and centre *both* in opposition to and in support of the industry.

An Economic Saviour?

The past twenty years have been difficult times for Canada's Atlantic and Pacific coasts. On the Atlantic coast, the moratorium on the northern cod fishery that was imposed in 1992 has devastated many communities, particularly in the province of Newfoundland and Labrador (Sider 2003). The impact of the collapse of this fishery on coastal communities is difficult to overstate. In addition to unemployment and household economic strain, research by Sinclair and colleagues (1999) on Newfoundland's Bonavista Peninsula uncovered a distressing level of despondency, disaffection with life chances, and apathy in the face of such dramatic individual- and community-level change. While new fisheries are being encouraged (such as crab and lobster fisheries), environmental concerns are limiting their size (Kennedy 2006; FRCC 2007). On the Pacific coast, efforts to reduce pressure on the salmon resource through the reduction of licences mean that the fleets of many former coastal centres have been dramatically reduced (Brown 2005). Moreover, British Columbia's coastal forestry sector, which was once the economic engine of the province, has a suffered a steep decline over the past twenty years (Hayter and Barnes 1997; Young 2008). The collapse of coastal forestry follows a too-familiar pattern, where environmental degradation and poor resource management pushed a seemingly boundless natural resource to a point that Marchak and colleagues (1999) describe as "falldown," where it becomes increasingly expensive to harvest and process forests that are of diminishing quality. Following the liberalization of forest policy implemented by the BC Liberal government in 2002-03, the forest sector has accelerated its retreat from coastal regions and concentrated employment and investment in Interior regions of the province (Young 2008).

These economic difficulties have led to significant demographic change in both regions. Table 2.1 shows population declines in selected communities from 1991 to 2006. Each of these communities is rural and mid-sized (with 1991 populations between 5,000 and 20,000), with the exception of Bella Coola, which had a 1991 population of less than 1,000 but played a major administrative and economic role in BC's sparsely populated Central Coast region. Moreover, these communities are not temporary or "instant" towns – they are well-established regional centres with deep roots in traditional fisheries, forestry, and other resource sectors. These losses have a psychological impact on many who stay, including a deep uncertainty about the future and the role of these communities in the broader economy (Young 2006a).

The encounter between the aquaculture industry and these economic and social circumstances is complex. The industry and its supporters have for some time promoted aquaculture as good news for coastal communities under stress. According to this argument, the industry offers a way to stabilize local economies through steady employment and infusions of investment

Table 2.1

Population changes in mid-sized Atlantic and Pacific communities in Canada, 1991-2006

	Community	Percentage change in population
Atlantic	Antigonish, NS	−14.0
	Miramichi, NB	−14.3
	Bonavista, NL	−18.1
	Marystown, NL	−19.3
Pacific	Bella Coola, BC	−54.2
	Port Alberni, BC	−5.2
	Port Hardy, BC	−24.8
	Prince Rupert, BC	−22.9

(e.g., CAIA 2006a). This view is shared by many coastal residents, businesses, and communities. The opposite view is also widely held, however – that aquaculture is in fact destabilizing already-fragile coastal economies. As we shall see, both views are emotionally and passionately articulated in a way that that leads us to argue that aquaculture has become a powerful but fluid metaphor, representing a myriad of hopes and fears under very stressful economic and environmental circumstances.

Local understandings of aquaculture are often difficult to assess. Public debate over aquaculture is dominated by the representatives of industry and environmentalist groups (see Part 2), while local voices receive notably less media or political attention. To examine local perspectives, we draw on transcripts of the government of British Columbia's Special Committee on Sustainable Aquaculture (SCSA) as a means of examining local understandings of aquaculture.[1] The SCSA conducted hearings in nineteen communities in coastal British Columbia, and all testimony has been made public as part of the official record of the Legislative Assembly of British Columbia.

There are several advantages and limitations to our use of SCSA transcripts to examine local understandings of aquaculture. The foremost limitation is that the committee restricted its hearings to British Columbia, and no comparable record exists for the Atlantic coast or other regions where aquaculture is practised. The second limitation is that local witnesses before the committee were self-selected, meaning that there are limits to the generalizability of findings to all coastal residents. In our view, however, these limitations are in some ways countered by the richness and breadth of the data presented in these transcripts as a window into local perceptions of the industry. The open-ended testimony, which includes the words of community leaders, local interest groups, and concerned citizens of both Aboriginal and non-Aboriginal backgrounds, provide us with access to people's thoughts, emotions, and explanations about aquaculture in their own words. These words give us a better understanding of the multiple meanings that aquaculture has in local settings. As such, these transcripts are a valuable tool for sociological analysis of the aquaculture controversy, for understanding how and why this particular issue has generated such strong responses.

Livelihood and Autonomy

The SCSA heard strong and passionate opinions both in favour of and in opposition to aquaculture. Interestingly, much of the testimony for and against aquaculture shows significant thematic overlap. Specifically, much of *both support of and opposition to* aquaculture is articulated with reference to two key themes: threats to local livelihoods, and/or the rejection of outside influence in local affairs. These thematic overlaps are worthy of careful consideration, as they suggest (along the lines of the argument in the previous chapter) that the aquaculture question is entangled in larger issues of economy, governance, and environment.

On the first point, both local supporters and opponents speak of how the industry relates to a perceived immediate threat to livelihood. Among supporters, this is usually expressed in what we will call a "continuity narrative" – that is, community members speaking to the SCSA in support of aquaculture often frame it as a means of bridging economic eras while maintaining community stability. In this view, aquaculture represents an opportunity to arrest disruption and change at family and community levels.

> The small communities up and down the coast, both native and non-native, rely on this industry to sustain their way of life, raising their children in their small communities as I did mine in Campbell River in its infancy. Now some of my children work in the aquaculture industry and are about to raise their children with these resource dollars. It is very easy for the naysayers to oppose everything that is proposed, but we the workers of B.C. would like to have a job right here at home to keep our families fed and clothed with good, clean resource jobs like these aquaculture jobs. In short, resources pay the bills – your wages, my wages, the schoolteacher's, the doctor's, the nurse's, and on and on – so let's help to make it happen and not further block this industry out. (Dennis Walker, Community Member, spoken at Campbell River, BC[2])

> I am very grateful for this farm that is here. As one family member here, it's given me a lot – to stand up for myself and be proud of myself and proud of this community for the positive things that it's given us, for the work. It's not just about work. It's about the people that I've come back to see. A lot of these people here I haven't seen for years and years ... It's nice to come

back home to this community ... It's about the farm. I do have positive things to say about this farm ... and I am grateful that [it is] here. (Wayne Starr, Member of the Kitasoo-Xai'xais First Nation, spoken at Klemtu, BC)

The continuity narrative is consistent with arguments made by the industry and governments that "we are witnessing the coming of age of a new industry and a new hope ... that will provide immediate and long-term economic stimulus in rural and coastal communities" (Newfoundland and Labrador 2007). It would be simplistic, however, to consider the local continuity narrative as motivated only by instrumental and economic considerations. Questions of livelihood run very deep in rural and coastal areas – these are not just about economics but are closely bound to issues of identity and sense of place (Sinclair et al. 1999; Young 2006a). It is perhaps for this reason that the continuity narrative is often entwined with a deep hostility towards perceived outside influence over the aquaculture industry. This hostility is directed in part at government, but particularly towards urban lifestyles and environmental activism.

We fear rightly that as a result of legislative initiatives like this [the SCSA], our jobs will end and we will have to move to the big, ugly city. Yes, God help us survive any more legislative fixes that end with politicos shaking hands and smiling to the world's elite while even more rural workers pack their bags.

I've worked ... at the fish plants ... When you take a moment to look across the table at a young, working single mom cutting a fillet there, with a returning post-secondary student beside her, and then you look to your left and your right to see another single mom and yet another ... When you do that, you get a bit of a shudder, when you think of those who would come in with flippant opinions of their industry ... that is to say, with their minds already made up. God save us from those kinds of people. (Bruce Lloyd, Community Member, spoken at Port McNeill, BC)

Most ... [workers] are not here for the paycheque, as is often found in other industries. They're here because they love the outdoors and the ocean. It's a hard job. They're away from their families. They can probably work in a plant somewhere pushing a button a lot easier. But they're here because they enjoy the outdoors, they enjoy the ocean, and they enjoy the community ... These are your real environmentalists. These are the people who care about the community. They care about what they're doing and where they want to be ... They're not sitting in Vancouver, driving around in their SUVs. (Dave Adams, Area Manager for Pan Fish, Inc., spoken at Port Hardy, BC)

As mentioned, local *opposition* to aquaculture is often articulated in notably similar terms, specifically with reference to livelihood and the imposition of outside interests in local spaces. But while the themes are similar, the logic and argument are rather different. The first difference is that oppositional narratives put much more emphasis on questions of environment. While the continuity narrative understands aquaculture as a means to stabilize communities in the face of change, a countervailing "disruption narrative" frames aquaculture as a direct threat and/or cause of the breakdown of traditional practices and livelihoods. In this narrative, environmental degradation is the primary cause of disruption on the coast, with economic hardships following environmental problems. As such, the environmental costs of aquaculture are understood as part of the problem, rather than as a potential solution.

> What's going to happen? If we wind up losing [the wild] salmon ...? A future for our children – this is a really important part. We've lost our logging, and it's going to be a long time before we can actually start going again ... With that, if we lose our [wild] salmon – everybody who works in the north is related to salmon, no matter what. You don't think you are. Everybody that lives up here is related to salmon – every company ... It's really important that you guys [the SCSA] understand that the decision you are making has to be a viable one. (Fred Seiler, Small Business Owner, spoken at Terrace, BC)

> I can tell you that people are going to be furious. The wild salmon of our area are the lifeblood, the soul, of our coast. If [salmon farms] are allowed here, they will tear apart our soul, and you will see a lot of angry people. (Greg Knox, Small Business Owner, spoken at Terrace, BC)

A second important difference between local narratives in support of and in opposition to aquaculture is that the latter often contain strong reactions against the suggestion of dependency. As previously discussed, local supporters of aquaculture articulate a deep hostility towards an outside, and particularly an urban-environmentalist, influence over the industry. While many local opponents share this hostility, their targets and rationale differ. As outlined in Chapter 1, resource regions in Canada have long suffered a relationship of dependency with urban centres, where local resources have been controlled and abused by distant governments and corporate actors (Marchak 1983; Matthews 1983). The legacy of dependency is a major point of reference for local opposition to the industry.

> While finfish farms ... are feeding consumers, they will starve the people of the Skeena [River]. While finfish farms ... will line the pockets of a wealthy

few, they will devastate the people of this watershed. While the resources, lifestyles, culture and history of the people of the Skeena slowly disappear with these developments, you [the SCSA] will be left with nothing but your mark in history as those that knowingly destroyed the lifeblood of this mighty river and its people, all in the name of greed. (Shannon McPhail, Community Member, spoken at Kispiox, BC)

I'm old enough now that I've seen a fair bit of things come into this country here. We get all kinds of directives from Victoria. We get blanketed with the same kinds of things up here that don't fit because the decisions are made in Victoria and Vancouver. Up here what we have are mountains and forests, and we've got fish and rivers. We don't have much of a forest left because we've had quite a few companies come in from different places and set up shop and "provide jobs." But it's all been quite short-term. It looks to me, from what I can understand with this fish farm business, that this is an employment opportunity. But it looks really shortsighted as well, so I just don't think it's a good idea. (Brian Larson, Community Member, spoken at Hazelton, BC)

The roots of this community include the value of meaningful work being more important than monetary wealth ... I am dismayed, then when it appears as though the [SCSA] wishes to build a "sustainable" aquaculture by sacrificing this lifestyle, the vocation and freedom to choose fishing wild salmon ...

We believe that the right way is our way. With the coming of the salmon farms the opportunity to choose fishing as a vocation and lifestyle is being eliminated quickly. (Lynne Belfry, Community Member, spoken at Sointula, BC)

In the Introduction, we talked about Leiss' (2001) notion of the labyrinth as a means of conceptualizing the complexities that arise when industries or firms misjudge what their actions *mean* to public and stakeholder groups that are increasingly knowledgeable, vocal, and politically influential. These quotations clearly show that local people are having multiple encounters with aquaculture, some positive and some negative, based in large part on issues of livelihood and locality.[3] The aquaculture industry often presents itself as being "the key to coastal revitalization" (BCSFA n.d.; NBSGA n.d.a). Some local people agree with this; others do not.

Drawing further on the SCSA transcripts, we argue that industry's enthusiasm to be seen as a saviour of coastal communities and livelihoods may have contributed to its stumbling into the labyrinth. Specifically, for many local people, the suggestion that aquaculture can save or revitalize coastal lifestyles immediately shines the spotlight on what is being devalued or lost.

If you love the fish, you've got to look after them. You don't just use them; you don't just take them and abuse them ... I hear a lot of talk about how fish farming is going to change this community [for the better]. I don't see it, to tell you the truth. I really don't ... Why are we allowing ourselves to get stuck in one place, put all our eggs in one basket and hope that fish farming is the answer? I don't see it that way. (Ken Innes, Member of the Gitxaala Nation, spoken at Kitkatla, BC)

Human spirits do not soar with nothing more than fenced-in livestock for human consumption only. Coastal B.C. is not paradise because the word "wild" is missing in our lexicon – quite the opposite. Just ask yourself: "What will make me" – meaning you – "happier: seas of floating netcages or legions of salmon returning to our rivers to spawn?" It won't be both. (Van Egan, Community Member, spoken at Campbell River, BC)

I heard someone allude [in this hearing] to fish farming as one of the [only] things that is left here in Clayoquot Sound. Now what are we going to do when all these things are destroyed? I have a very difficult time with it. You know, I could look into the eyes of each one of these people that work in the fish farm industry and say: "I cannot blame you for wanting to have a job." I can't – how can we? – but we do have to be careful with what we have left. That is going to be the future of these little ones, our people – Tla-o-qui-aht – and all the coastal people that live along here. (Joe Martin, Member of the Tla-o-qui-aht Nation, spoken at Tofino, BC)

By casting aquaculture as an economic saviour and problem solver for troubled regions, industry enthusiasts have unwittingly married the industry to the problem of coastal economic and environmental hardship. Industry advocates often make what we would term "rational-instrumental" arguments for aquaculture development. From this perspective, saving coastal communities is primarily an economic task. As can be seen in the following quotations, the transition from capture fishers to aquaculture is presented as rational and reasonable; one is a sunset industry, the other a promise of sunrise. Moreover, the transition is assumed to be smooth and orderly – coastal traditions and culture can be preserved by the sector's capacity to pick up the employment slack. The first quotation is from the New Brunswick Salmon Growers Association, and the second is from federal Office of the Commissioner for Aquaculture Development:

For many years, the traditional fisheries were not only an important contributor to New Brunswick's economy but they were also an integral part of the cultural identity of its people. As a result, the declines in the wild fish stocks brought challenges to the province's economy as well as to its unique cultural heritage. In the face of these challenges, the growth of the salmon farming industry has contributed greatly to an economic and cultural revitalization of New Brunswick ... In addition to providing about $90 million in wages, the jobs have reawakened the pride within the people of New Brunswick for their heritage – and their ability to harvest the ocean's bounty to feed the world. (NBSGA n.d.a)

At a time when Canadian fisheries have suffered from the reduction of the Pacific salmon and Atlantic groundfish stocks, the enhanced commercial harvest of shellfish has emerged to augment the production and value of the fishery. Today, however, the fishery is largely dependent on three species – crab (especially snow crab), shrimp, and lobsters, and the decline of any one of these stocks would have significant consequences for the sector. Moreover, uncertainty surrounds the status of the shellfish stocks and future harvest tonnages cannot be considered secure. In the absence of growth through aquaculture, many of Canada's coastal and rural communities face an uncertain future. (Bastien 2004, 8)

The argument that aquaculture can save coastal communities is in some ways politically expedient, as it highlights the upside of aquaculture development for investors, governments, and communities. On the other hand, it is proving to be politically suspect at the local level. Perhaps a wiser strategy among industry proponents would be to position aquaculture as *part* of a broader diversification of coastal economies. In reality, this is closer to the actual state of affairs, given the industry's current small stature and limited geographic presence on both Pacific and Atlantic coasts. Some coastal residents are committed to aquaculture as a way of preserving coastal living. For others, the argument that aquaculture is "the key to coastal revitalization" (BCSFA n.d.) evokes the problem more than the solution.

The Aboriginal Rights Question

This is the message from our people ... take it back and say that we, the First Nations people, have spoken. We have spoken. We come to you to tell you. We don't want these farms – 110 percent. If [they] ever come in, we as the Gitxsan people are going to come down and destroy [them]. (Roy Wilson, Member of the Gitxsan First Nation, spoken at Hazelton, BC)

We're sick and tired of this, and some of our people are ready to go to war. I've had three people come up to me and say: "I'll throw the first Molotov cocktail." This is what it's getting to. (Arthur Dick, Hereditary Chief, Namgis First Nation, spoken at Alert Bay, BC)

One of the most significant developments in Canadian politics over the past thirty years has been the re-emergence of the Aboriginal rights movement as a major legal and moral force – one that is particularly strong on questions of environment and resource management. The Aboriginal rights movement is the oldest social movement in Canada, but was strongly repressed by federal and provincial governments until the 1960s (cf. Harris 2002). Until 1951, federal law prohibited Aboriginal peoples from self-organizing in pursuit of land claims and other rights. Prior to 1960, Status Indians were denied the federal vote unless they surrendered treaty rights and removed themselves from the official register of the *Indian Act.*

Woolford (2005) argues that the elimination of these restrictions was the first step on a long road for the contemporary Aboriginal rights movement. Following enfranchisement, the governments of Prime Ministers Lester Pearson (1963-68) and Pierre Trudeau (1968-79, 1980-84) were very active on Aboriginal issues. The first key development occurred in 1965, when the Pearson government introduced Bill C-123, which established an arm's-length Indian Claims Commission (ICC) that included Aboriginal representation and that committed to funding Aboriginal groups to research and present land claims (Woolford 2005, 81). The second major development of this era was the release in 1969 of the government of Canada's "White Paper on Indian Affairs," officially the *Statement of the Government of Canada on Indian Policy.* The White Paper, which was spearheaded by Indian Affairs Minister and future Prime Minister Jean Chrétien, envisioned sweeping changes that included repeal of the *Indian Act,* the transformation of reserve lands into fee-simple property, and the end of taxation exemptions (Woolford 2005, 59). The White Paper was intended to be a progressive stance on Aboriginal issues, rectifying previous rights imbalances through the granting of full citizenship to Aboriginal people. According to Tennant (1990, 139), it "was rooted in the longstanding small 'L' liberal ideological view that individual Indians desired to be and should be assimilated as equals into the larger Canadian society. Unfortunately this view ignored the attachment of Indians to their ancestral communities and tribal groups." Indeed, the White Paper stands as a pivotal moment in the Aboriginal rights movement, as it cemented a commitment to group rights as the focus of activism – a stance that presents a strong challenge to liberal democratic (and capitalist) conceptions of rights as held by discrete individuals and organizations (Ignatieff 2000).

In the years following the White Paper, Aboriginal groups in Canada have increasingly turned to the courts. For this discussion, the most significant of these court actions are the *Calder* case (1973), the *Sparrow* decision (1990), and the *Delgamuukw* decision (1997).[4] The case of *Calder v. British Columbia* involved a suit brought by the Nisga'a people in northwestern British Columbia against the province, "arguing that Nisga'a title to their lands preceded

British sovereignty, that this title had never been extinguished, and that such title continued to exist as a legal right" (Woolford 2005, 60). The *Calder* case was based on the fact that the vast majority of the province of British Columbia has never been ceded through treaty (except for the northeastern corner of the province, which is covered by Treaty 8, and the legally murky and still disputed "Douglas Treaties" on Vancouver Island). While *Calder* was defeated on a technicality at the Supreme Court of Canada, it spurred significant changes in government strategies for addressing First Nations grievances. According to Tennant (1990), the *Calder* case was pivotal in convincing the Trudeau government that the best way forward was to negotiate directly with Aboriginal groups, thus abandoning the liberal foundations of the White Paper (see also Sanders 1990).

The case of *R. v. Sparrow* involved a charge brought by the government of Canada against Ronald Sparrow, a member of the Musqueam Band in southwestern British Columbia, for fishing using illegal equipment. Sparrow's defence, which was successful, was that the regulation of fishing gear infringed on his Aboriginal right to fish for ceremonial, social, and sustenance purposes (Avio 1994). The case set a strong precedent for resource management issues, as it established that regulatory laws (such as the federal *Fisheries Act*) did not automatically override or extinguish Aboriginal rights. The *Sparrow* case has strong parallels to *R. v. Marshall*[5] (1999), in which the Supreme Court of Canada ruled that Donald Marshall Jr., a Mi'kmaq from the Membertou Band on Cape Breton Island, Nova Scotia, had a treaty-protected right to fish and sell eels, a right that superseded federal licensing and regulation (Coates 2000).

To date, the most important court case on Aboriginal rights is *Delgamuukw v. British Columbia*, a monstrously complicated suit that wound its way to the Supreme Court of Canada over fourteen years (it was launched in BC Supreme Court in 1984 and decided in the Supreme Court of Canada in 1997). Foremost, *Delgamuukw* involved a claim by the Gitxsan and Wet'suwet'en nations to 58,000 square kilometres of land in northwestern British Columbia based on traditional occupation and usage of the territory. Although the Supreme Court of Canada refrained from ruling on the land claim itself, the ruling made several acknowledgments of Aboriginal rights that have had profound ramifications for governments and resource firms across the country. Primary among these is "the specification that Aboriginal title is a pre-existing right to the land based on occupancy and that it does not directly correspond to conceptions of ownership held in British common law or French civil law" (Woolford 2005, 64). The Supreme Court also ruled that Aboriginal title is communally rather than individually held, further entrenching group rights as the primary frame of reference for Aboriginal rights. Finally, the Supreme Court decision stressed the responsibility of governments to negotiate final resolutions to questions of rights and title. In the interim, governments were to engage in meaningful consultation with Aboriginal groups on policies and projects that potentially conflicted with Aboriginal rights and title. This consultation was to be conducted according to the following principles: "That Aboriginal people whose title is being infringed should be consulted concerning the infringement, that there should be as little infringement as is necessary to accomplish the government's goal, and that fair compensation should be paid for any infringement" (Dacks 2002, 241).

The *Delgamuukw* decision in particular has provided Aboriginal groups across Canada with new tools for both protest and participation (Joffe 2000). The Supreme Court's acknowledgment of prior usage-based rights, in combination with its edict on consultation, has spurred significant changes in corporate and government approaches to industrial development and resource management. To be clear, *Delgamuukw* has not revolutionized treaty negotiations in Canada, which continue at a snail's pace in all regions of the country (Dacks 2002). It has, however, given significant weight to both Aboriginal protest and desire for meaningful participation in development even *outside of treaty settlements and negotiations.* Indeed, in the years following *Delgamuukw,* both governments and industry have demonstrated uncertainty on how to proceed. The business lobby, in particular, feared a situation where projects and operations would be stopped cold at the insistence of a First Nations group with claims to the territory (Woolford 2005). Other fears were raised that "if we follow the Supreme Court's proprietary view of Aboriginal title, we are likely to establish a racially-based 'rentier society,' in which resource rents no longer belong to [citizens] at large, through their provincial government, but instead to particular First Nations" (Scarfe 2000, 50).

Chaos has not followed the *Delgamuukw* decision, but the relationships among industry, government, and individual First Nations have changed. Given the legal force of these court decisions, provincial governments and resource firms in particular have developed a strong aversion to Aboriginal protest (Hayter 2003). In British Columbia, where Aboriginal rights have never been defined by treaty, there have been several high-profile cases in the late 1990s and early 2000s where resource firms have abandoned long-standing operations once confronted with actual or even threatened protest, including forestry operations on the vast Central Coast Region and on Haida Gwaii/Queen Charlotte Islands (Ramsay 2005; Young 2006a, 161). At the same time, governments and private firms have moved to form partnerships with Aboriginal groups (and vice versa) as a means of moving forward in the absence of formal treaty agreements (Parsons and Prest 2003; Matthews and Young 2005). Recent court cases have, if anything, slowed the already-glacial pace of formal processes such as the Indian Claims Commission and the British Columbia Treaty Process, in part because the federal government will not negotiate with parties that are engaged in litigation, and in part because court rulings to date have only encouraged negotiation without "compelling governments to change their offers" (Dacks 2002, 242). In this context, the benefits of partnership for private enterprises are significant, with partnership being a means to secure access to resources and acting as a buttress against environmentalist protest. These partnerships are controversial among Aboriginal people, however, and are a significant source of conflict within and across First Nations groups and communities (see below).

The primary effect of the Aboriginal rights movement, and particularly of the legal activism of the past several decades, has been to greatly empower *individual* Aboriginal groups with respect to development issues in traditional territories. It must be stressed that much of this empowerment is unofficial, stemming more from the threat of protest and further legal action than from official codified rules, which are the realm of slow-moving treaty and claims processes. It must also be stressed that governments and firms are still in many cases willing to force projects through despite Aboriginal opposition (cf. Peterson 2004). Nevertheless, individual First Nations groups are now more capable than ever of using protest and partnership to strongly affect the types of development that occur in traditional territories (Matthews and Young 2005).

This presents a complicated political landscape to industry, particularly the aquaculture industry, which is both highly controversial and a new claimant to coastal territories and resources. The empowerment of individual First Nations means that the "Aboriginal question" in resource regions is perhaps better cast as "First Nations questions," where individual First Nations are asserting their own views of which activities are environmentally and culturally acceptable on traditional territories. *Delgamuukw* has not led to the "rentier society" feared by Scarfe (2000), but rather to a situation where governments and firms are increasingly required to deal with multiple Aboriginal groups – to make a moral and material case for resource development on a Nation by Nation basis.

Livelihood, Colonialism, Rights

In many Aboriginal communities, the moral and material case for aquaculture development has clearly not been made, and both the industry and senior governments are frequent targets of harsh criticism. As will be seen below, these criticisms are complex and profound, and touch on a range of established grievances and fears for the future. But while Aboriginal opposition to salmon aquaculture is widespread, it is neither uniform nor universal. In some cases, aquaculture companies have taken a strong lead in forging agreements and partnerships with Aboriginal communities. The best-known example involves the agreement between Marine Harvest Canada Inc. and the Kitasoo-Xai'xais First Nation at the community of Klemtu on British Columbia's Central Coast. While the details of the agreement are secret, it involves a measure of community control over siting, guarantees of local employment, improved environmental monitoring, and a commitment to process harvest in the community (Larry Greba, testimony to the SCSA, Klemtu, BC). In 2002, the western Vancouver Island First Nations community of Ahousaht signed a similar agreement with Pacific National Aquaculture (Vodden and Kuecks 2003), as did the northwestern community of Kitkatla with Pan Fish Canada in 2003. (The expansion of salmon aquaculture to Kitkatla is in doubt, however, due to strong opposition from neighbouring Aboriginal groups.)

Fieldwork conducted by Schreiber and Newell (2006a) in Ahousaht during the time of the agreement with Pacific National Aquaculture shows how difficult the decision was for the community and its leadership. Although the BC Salmon Farmers Association trumpets Ahousaht as a case of a community that formerly opposed aquaculture but "now believes that mutual respect and harmonization with the industry is [sic] the best way to ensure that resources and the environment are protected" (BCSFA 2005a), Schreiber and Newell argue that this conclusion is too simplistic. Upon careful reading of communications between First Nations groups, industry, and government, they conclude:

> These communications deal both with mundane matters of fish farming practice and with still unfulfilled Aboriginal demands to resolve underlying questions about land and jurisdiction over resources. The policies of First Nations regarding fish farming, even when they are explicitly defined, are therefore the outcome of interactions and responses that cannot be categorized as simple acceptance (understood as forward-looking, or modern) or opposition (understood as clinging to the past, or traditional). Instead, the strategies employed by First Nations to deal with salmon farms ... are attempts to resolve the tension between, on the one hand, wanting to influence decisions being made over fish farming and, on the other, standing by what First Nations people have always held to be their inalienable rights to build a future on their own terms. (Schreiber and Newell 2006a, 87)

The point is not that First Nations support for partnership with salmon aquaculture firms is weak or contingent, but that it too is complex. Aquaculture is entangled with multiple contentious economic and environmental issues. We will therefore argue below that First Nations' acceptance or tolerance of aquaculture is as complex as the stances taken against the industry.

Figure 2.1 summarizes some of the major themes that were articulated by First Nations speakers at the SCSA in support of and in opposition to aquaculture development. The structure of this figure merits some explanation. First, the figure highlights the fact that the two major scientific narratives against aquaculture development – environmental integrity and human

Figure 2.1

Key themes in First Nations testimony to British Columbia's Special Committee on Sustainable Aquaculture in opposition to and in support of aquaculture

Opposition	Support
Environment	–
Health	–
Livelihood (subsistence)	Livelihood (employment)
Colonialism (violence)	Colonialism (dependency)
Rights	Rights

health – are also part of Aboriginal objections to the industry. With respect to the environment, the overwhelming concern was the perceived threat that industrialized and mechanized aquaculture development poses to wild species and habitat:

> We have all the technology geared up to destroy what keeps us going. We as human beings, as First Nations people, depend on that salmon. We are not the only ones that depend on it. All the animals in the forest – the trees, the birds, the bears, everything – depend on that salmon. Once that salmon is gone, where are we going to be? ... My children and grandchildren are not going to have the luxury of seeing a wild salmon, at the rate we're going. They'll probably only see farmed salmon. That is exactly what we don't want here. We don't need that. (Peter Siwallace, Hereditary Chief, Nuxalk Nation, spoken at Bella Coola, BC)

With respect to health, it is important to note that, generally speaking, both individual and community health are important concepts in First Nations culture (Waldram et al. 2006). It is well known among epidemiologists and public health officials in Canada that Aboriginal communities suffer some of the poorest overall health conditions in the country (MacMillan et al. 1996). The general consensus is that poverty and lack of access to traditional foods has pushed many Aboriginal people towards poor diet (Tookenay 1996; Newbold 1998). Moreover, some researchers claim that the effects of malnutrition are exaggerated in First Nations populations due to specific genetic vulnerabilities, particularly since traditional diets in many hunter-gatherer societies contained very little carbohydrate (Robinson et al. 1995; Hegele et al. 1999). The results of this include extremely high rates of diabetes, cardiovascular diseases, and some cancers (Young et al. 2000; Anand et al. 2001). Given these challenges, it is not be surprising that aquaculture should be interpreted through a health lens. As discussed in the Introduction, human health is a key axis of the broader controversy over aquaculture, particularly in terms of possible contamination of aquaculture products with industrial chemicals such as PCBs. Although the debate over this type of contamination is playing out primarily in scientific journals and expert language, the notions of taint and disease clearly have strong local resonance as well:

> I think the big question was, at the beginning: do we as Kwakiutl people here approve of having fish farms in our area? Right from the very outset, the answer was no ... We've had so many younger people, the next generation down from me, pass away just recently. My nephew just passed away the other day. Another friend of mine passed away about four days ago. Another friend of mine just called me tonight – my best buddy. He says to me: "Stan, I've got bad news. I've got cancer" ...
>
> Our people are dying because of a lack of access to our [traditional] food, and that's all it boils down to. I've had so many friends who have passed away in the last seven or eight months because of cancer, diabetes, you name it. I think the Royal Commission on Aboriginal Peoples already said that a long time ago. It appears to me that nobody's listening. The answer is still a big, flat no. We don't want fish farms here. We've got one of the most pristine areas in the province right here, and we're being ruined. (Stanley Hunt, Member of the Namgis First Nation, spoken at Alert Bay, BC)

> [My father] is absolutely opposed to any [salmon] farm on our tribal land or our tribal territories, or our tribal waters. Also, [my father] wants to know why, with all the research done and all the government spending on preventing cancer in Canadians, they would subject the First Nations to the possibility of getting cancer from these fish? They have toxins, they have cancer-causing agents in them. Why would they subject the First Nations to this sort of thing – literally, psychologically, emotionally, physically? ... Establishing fish farms in our tribal territories and letting one farmed fish out is equivalent to letting TB [tuberculosis] out in the world. How do you contain it once it's gone? (Raija Reid, Member of the Heiltsuk First Nation, spoken at Bella Bella, BC)

Figure 2.1 also illustrates our finding that Aboriginal opposition to and support of aquaculture development often reflect "paired themes" or mirrored arguments. Many statements to the SCSA from First Nations people, both for and against aquaculture development, demonstrate thematic consistencies even when they are substantively very different. We observe these consistencies across the themes of livelihood, colonialism, and rights, which are each raised in opposition to and in support of aquaculture development. We stress, however, that this is not a case of point/counterpoint. For instance, as we will see shortly, the theme of colonialism plays a strong role in oppositional stances to aquaculture articulated to the SCSA. The association of aquaculture with the legacies of colonialism runs deep and is extreme in some cases, as some First Nations people describe aquaculture using language that includes terms such as cultural and ethnic genocide (see also Schreiber and Newell 2006b). Conversely, some Aboriginal supporters of aquaculture development describe the industry as a means of breaking with a colonial past (in the form of state dependency and dominance over local affairs). Although we consider these arguments to be related, they are not opposed to one another, in that they are not formulaic, mobilized to counter other claims, or intended to make a public point. Rather, these are candid articulations that reflect some of the common themes of contemporary Aboriginal life in Canada, including the struggle for self-determination, reconciliation, and cultural continuity (cf. Warry 1998; Woolford 2005).

We argue, therefore, that understanding the context in which the paired themes outlined in Figure 2.1 are evoked is key to understanding why aquaculture remains intensely controversial at the local level. That is, the testimony before the SCSA gives an indication of how local understandings of aquaculture are expansive, incorporating other issues and tensions that predate the industry. In the minds of many Aboriginal leaders and community members, the aquaculture question is about much more than industrial development, much more than environmental and health risks, and much more than the prospect of a "rentier" relationship with government and industry.

The first thematic pairing in Figure 2.1 takes us back to the question of livelihood. As discussed in the previous section, the notion of livelihood resonates strongly among both Aboriginal and non-Aboriginal coastal residents: whether aquaculture is understood as a continuation or a disruption of livelihood is a major fault line in local understandings of the industry. For many First Nations communities, however, questions of livelihood run deeper still, and are strongly intertwined with issues of culture and environment. Coastal First Nations peoples have long legacies of marine use that predate colonial and capitalist economies. This means that the connection between aquatic resources and livelihood touches on themes of tradition and continuity that do not apply (or at least not as intensely) to non-Aboriginal coastal residents. Thus, for several Aboriginal speakers at the SCSA hearings, aquaculture's perceived threat to livelihood is understood primarily with reference to the subsistence fishery, which is highly valued in both economic and cultural terms.

> We are people closely connected to the land. We're closely connected to each other. The salmon resource is an important part of that. We have 90 percent unemployment. Most of the people here are impoverished. It's not a luxury to have the salmon returning. It is essential for some people in

order for survival. If we don't have the salmon in jars, in the pantry or in the freezer, people will go hungry. It's also an intimate part of the culture. I think about the buffalo and what it meant to the people of the plains and that culture. I do not want to be part of a system that says to our grandchildren: "There used to be a salmon culture in the Skeena River, and now there no longer is." (Doug Donaldson, Member of the Gitxsan First Nation and City Councillor, spoken at Hazelton, BC)

We've been here for 10,000 years. The main sustenance of our people is salmon ... You talk about farmed salmon. You're not only talking about farmed salmon; you're talking about the grizzly bear, the ecosystem and the birds. It's not only the salmon that is involved here; it's a huge ecosystem. It's almost like the government is slowly destroying the First Nations people. It started way back when they tried to get rid of our feasts, putting us in residential schools, taking away our culture and our language. Now they are going to try taking away our salmon. It isn't going to work. We will not allow it. (Yvonne Lattie, Hereditary Chief, Gitxsan First Nation, spoken at Hazelton, BC)

On the other side of the ledger, Aboriginal supporters of aquaculture also talk about livelihood as being linked to culture and tradition. In this case, however, aquaculture is understood as a means of recapturing a measure of prosperity, independence, and self-determination. The issue of employment is central in this narrative, but it is important to note that employment is rarely invoked as a stand-alone reason for supporting aquaculture development. For instance, in the first quotation presented below, the issue of employment is intertwined with problems of dependency and state patronage. In the second quotation, employment is linked with the value of community and place.

So many, the federal government included, have been holding the hands of my people for so long, and it's a shame. I'll tell you why I say this. Indian Affairs, when they first came out here giving out welfare – I forget what they called it. They called it relief. They came out here and had a big meeting [and said] "Oh, you don't have to work anymore. We'll give you all the money you want. All you have to do is go to the store and get some food."

But our people were fishermen. They were hunters. You know what one of the elders said at that time? He said: "You have just spoiled the Indian." That is what has happened ... We want to stop holding the hands of people who want to put food in our mouths. Our people want to do it themselves, and they want to do it with aquaculture that we believe is sustainable and environmentally sound. (Vern Jackson, Member of the Gitxaala Nation, spoken at Kitkatla, BC)

I'm very proud to be associated with salmon farming like Creative Salmon. It means a lot to our community not just for employment opportunities but for many things in the area. Today I support Creative Salmon, something that I wouldn't have done 15 years ago. Why? Because I fought hard on zero impact to the environment. It's something that I was good at. I worked with the environmental groups ...

Why the change now? Well, I paid a hell of a price for what I did in those days ... We closed down the logging industry, pretty much, in Clayoquot Sound. We've done other things [too] that left people with no employment opportunities ... I wonder: could it be any different if I had done things differently 25 years ago? Would my family still be here? I hope my two young sons won't leave here but will live here like I have. (Moses Martin, Chief Councillor, Tla-o-qui-aht, spoken at Tofino, BC)

The second thematic pairing in Figure 2.1 has to do with colonialism and independence. First Nations people and leadership continue to have a very complicated relationship with senior governments in Canada. Since the passage of the *Indian Act* by the government of Canada in 1876, Aboriginal affairs and reserve lands have been under federal jurisdiction. Among the more controversial legacies of the *Indian Act* has been the designation of Status Indians as wards of the state, exempt from certain obligations of full citizenship (such as taxation) but subject to a much stronger federal hand in community governance (cf. Cairns 2000). Both First Nations supporters and opponents of aquaculture speak to this legacy. For opponents, objections are generally framed in the language of colonialism – of the imposition of unwanted industry by aggressive, belligerent, and distant governments that are either ignorant of or hostile to First Nations governments and people. The language used in articulating this understanding is often highly charged, most notably in the use of the weighty term "genocide."

If we say no [to salmon farms], that's what we mean. For the government to proceed with their economic development plans would be an act of contempt against all First Nations in the northwest. The destruction of our wild salmon stocks is just another act of cultural genocide. You have no respect for the northwest, for aboriginal people. Yet all of our resources keep the lights on in Vancouver and Victoria. And when you're done with us – we're a desolate community – no resources to sustain ourselves. It is poor management and irresponsible management of a government when it's going to do that. (Ralph Michele, Hereditary Chief, Gitxsan First Nation, spoken at Hazelton, BC)

We're not here today to be complicit or an accessory to the fact of marine genocide ... Also, there's ethnic genocide ... It's very humiliating to sit here, equivalent to Prince Charles, begging for some of his land back in England. I don't say that lightly. I was thinking about my ancestors while some of the presenters [to the SCSA] were making their presentations ... The papal bull of 1455, Pope Nicholas V, another papal bull in 1493, Pope Alexander VI – a couple of paragraphs into that is where the licence to invade the Americas was given. Some of the wording that I copied from those ancient papal bulls that haven't been rescinded I'll read into the record: "To vanquish and subdue all us nations of free Americas and a licence to commit genocide." (Russell Kwakseestahla, Hereditary Chief, Laich-Kwil-Ta, spoken at Campbell River, BC)

As evidenced by these quotations, the link that is drawn between colonialism and aquaculture is strong but not necessarily direct. That is, in these cases, complaints against aquaculture development are articulated with reference to other longstanding grievances, particularly those associated with treaties, rights, and environmental degradation. The references to genocide and violence indicate just how serious these linkages are, however. In this context, it would be a grave oversimplification to consider aquaculture as just the latest in a list of Aboriginal complaints against industry and government. Rather, much more seriously, for many First Nations people, it appears that aquaculture has become symbolic of, or at least embedded in, some of the most explosive and fundamental historical sources of Aboriginal anger and frustration.

I have made this observation before: when the United States wanted to bring the aboriginal people to their knees, they opened season on the buffalo and destroyed millions of them. My observation here is that as long as there are trees out on the territories and there are fish in the rivers, they can't bring the aboriginal people to their knees. It's almost as if – and I'm not accusing anybody – it's been mismanaged to the point to take the wealth away from native people so that they will go into abject poverty and accept a lower [treaty] deal than it should be. (Ron Harris, Member of the Gitwangak First Nation, spoken at Kitwanga, BC)

Smallpox. We almost got wiped out by it. There were 3,500 of us before the [signing of the Douglas] Treaty. By the time of the treaty there were 135. You've got to think about me, and I've got to think about my grandkids. I've got to think about them first. I should have a future on my own land, no matter what anybody says, no matter what you say. It makes me angry if I have to come and talk this way – I shouldn't be – because we never got properly consulted – never. I'm angry. I have a reason to be, because I'm looking forward to a better future ... We opposed the [proposed offshore] oil and gas [development] ... We stopped that because we had to. Now we're opposing this 100 percent. (Rupert Wilson, Band Councillor, Kwakiutl First Nation, spoken at Port Hardy, BC)

Nothing is [here] anymore. So it goes for the rest of our land. It's been depleted of everything. As soon as – what would you say now? – the Europeans came amongst us, maybe 200 years ago, it's been a downhill slide since then ... White man's business started. They started to make money, and they said: "Well, we need more land. Move the Indians out." The missionaries were told to get rid of the Indians but to save the man. That means: "Take the Indian out of the man, but save him so he can work for us. Then we ran into places called residential schools. There are not many amongst us that haven't been touched by that ... You said to yourselves that all these Indians running around here didn't need all this land. You started inventing ways to say: "Well, we own the land. You don't own any land. This is Crown land." Now, can any of you tell me how on God's green earth a crown got onto my land? ... Our answer to fish farming is: absolutely not. No. Negative. None. (Laurence Helin, Elder, Lax Kw'alaams Band, spoken at Prince Rupert, BC)

The theme of colonialism is also expressed by some Aboriginal supporters of aquaculture development, although in more limited terms. While opponents commonly discuss aquaculture in the context of colonial violence (against environment, culture, and populations), supporters generally speak of it in terms of a struggle against dependency. In these cases, aquaculture is considered a means of breaking dependencies that have arisen through state patronage or more generally through contact with outside influences. Importantly, this includes both economic dependency (specifically social assistance) and the physical dependencies of addictions and poor health.

You're looking at communities that are 90 percent unemployed, band offices giving out welfare cheques, and watching your members going and doing other things. You know, alcohol is just devastating in our communities ... This isn't just about growing farmed salmon. It's about people living in communities like Kitasoo – isolated. Why would you want to put yourself in isolation? This is home for most people here. This is where they grew up, and they're comfortable here ... I love this place. It's my home and always will be my home. When there is so much unemployment – I guess the hard thing you see in life is that there's always a preoccupation with alcohol and drugs. (Ben Robinson, Member of the Kitasoo-Xai'xais First Nation, spoken at Klemtu, BC)

I know in our community – you know what? – we're basically sick and tired of handouts, and I think we're a lot more open to hand-ups. We want to get up, and we want to earn our own way. I think in a lot of other communities in the future, in aboriginal communities, you're going to see that starting to happen. A lot of aboriginal communities can stand up and say "You know what? We don't agree with all these things [development]," but at the same time you have to ask them: "Do you agree with all the suicides in your community? Do you agree with foster care? Do you agree with all these symptoms that are in your community?" (Ted Williams, Member of the Cowichan Tribes, spoken at Victoria, BC)

Kitasoo [Klemtu, BC] is a fish community – wild fish. We have a fish cannery over there ... where everybody worked there in season – be out in a fish boat or a fish packer or working in the plant. Everybody worked. That closed down. The wild salmon closed down ... Who came to our aid? Nobody ... We have done wonders since we got into partnership with Marine Harvest. Our young people are coming out and going shopping. They never used to do that before. They go to Port Hardy. Get lots of grub for their families and other things. Before that, they just relied on welfare, which we don't want. (Archie Robinson, Hereditary Chief, Kitasoo-Xai'xais First Nation, spoken at Klemtu, BC)

The final thematic pairing outlined in Figure 2.1 is that of rights. As suggested at the outset of this section, the issue of rights is increasingly at the fore of resource and rural development in this country. Indeed, rights claims are central to Aboriginal stances for and against aquaculture development as expressed to the SCSA. For opponents, the question of rights is relatively simple: governments have no authority to allow expansion of the industry into Aboriginal traditional territories. From this perspective, opposition to aquaculture is part of a larger struggle for self-determination and authority over development.

[This] really is a slap in the face to us people who belong here. We've never ceded, signed any documents or gone to war for our territory ... I'm here because I know that in my lifetime nothing's going to change. But I'm fighting like hell for my grandchildren to have a life – not something that's controlled by the world trade order. (Arthur Dick, Hereditary Chief, Namgis First Nation, spoken at Alert Bay, BC)

We've never given anybody rights to this territory. We've made this statement over and over for hundreds of years now, since you people came from wherever you came from. Our people made a stand in Ista in 1995 against the logging corporation. Our people made a stand against the hatchery in Ocean Falls. This is not a territorial issue. It's our human rights issue, because we are the salmon people. (Deric Snow, Hereditary Chief, Nuxalk Nation, spoken at Bella Coola, BC)

We have nothing here, but we are happy. You come here and impose on us, and we are unhappy. It's not right. .. It's not right when a foreign government comes in here and imposes all this stuff on us ... It's always anything that's put in place is to destroy in this valley ... If you could come back here later on after this is done and say how good it is, then I'll bow my head to you. But if it destroys our environment, why did you impose it on us? Why? (Cecil Moody, Member of the Nuxalk Nation, Bella Coola, BC)

The centrality of rights claims in oppositional narratives is not surprising. As alluded to in the quotation from Chief Snow above, First Nations engagement in British Columbia's "war in the woods" and other recent resource conflicts across Canada have been based on the assertion of prior rights against government sanction of corporate development (cf. Coates 2000; Jackson and Curry 2004). What is often not well appreciated, however, is that rights discourses run just as strongly through narratives *in support of aquaculture development* as through those that oppose it. This is significant in several respects. First, it belies any facile assumptions that those First Nations groups that pursue aquaculture development are supportive of industry or government actions per se. As can be readily seen in the quotations that follow, Aboriginal support for aquaculture cannot be assumed to mean endorsement or static acceptance of industry or government actions. Second, the articulation of rights discourse here suggests that support of aquaculture involves the *active assertion* of rights claims. In the language presented below, the decision to pursue aquaculture is not a compromise. While there is partnership involved, and the aquaculture firms involved in these partnerships are praised by the speakers, there is also a sense of Aboriginal unilateralism and assertiveness in how these agreements are framed. Clearly, participation in the aquaculture industry is not understood as giving anything up in terms of rights and authority.

The Gitxaala Nation has never yielded nor surrendered nor withdrawn aboriginal rights and title, and we'll continue to assert our interest and to exercise our nation's aboriginal rights over the three [proposed aquaculture] sites ... Gitxaala Nation is committed to the protection of aboriginal rights and title of our territories. Through this economic development the Gitxaala Nation will exercise its inherent right of self-government, self-determination, and through Pan Fish, provide an industry that is viable, sustainable and environmentally sound ...

Our community has always been devastated. We have always gone through such hardship with a high unemployment rate. This community – which I've just mentioned that we've inhabited from time immemorial – was once a thriving community within the salmon industry. Today when you people walked up here, did you count how many gill-netters we have? Did you count how many seine boats? Just about nothing. (Vern Jackson, Member of the Gitxaala Nation, spoken at Kitkatla, BC)

In our communities, we continually suffer approximately 80-percent un-employment year-round. Once again, we First Nations have been left sitting on the beach as activity takes place all around us. We did not fare well with the logging. We did not fare well with mining. Our wild stocks are down ... We're just sitting here watching as everything else takes place around us. This alternative, I'm hoping, will allow us to finally get a grip on and become masters of our own destiny. We need to look outside the box. (Ric Miller, Member of the Kitsumkalum Band, spoken at Terrace, BC)

We have not given up any rights to Gitxaala territory – not one inch. The province hasn't bombed us out of here, and they haven't taken the land from us. We haven't sold it to them, we haven't given it to them, and we refuse to give up an inch to the province. We haven't given up an inch to the federal government, and we're not going to start today ...

The number one concern that we heard [from our community] was the environmental process. Those who had signed on the agreement basically said that the environmental concern is in the contract, that if Omega [Salmon] ... didn't live up to the environmental concerns, Gitxaala had the first right to tell Omega Salmon to go away, to get out of Gitxaala territory ...

Our people grew up on the water. We're looking at forestry, and our people don't really want to touch the power saw. They don't want to touch the tree. What they feel comfortable with is being on the water. That is where they grew up. That is where we grew up. That is where we get our food source from. The most comfort for our people is on the water. (Clifford White, Chief Councillor, Gitxaala Nation, spoken at Kitkatla, BC)

In summary, the controversy over aquaculture is being strongly framed by Aboriginal issues and stances, particularly in British Columbia and in reference to salmon aquaculture. But how to make sense of all this? We have seen that while the Aboriginal rights movement in Canada has not enjoyed great success in accelerating formal treaty processes, the legal and on-the-ground activism of the past thirty years have greatly empowered individual First Nations groups. The legacy of court battles such as *Calder* and *Delga-muukw,* as well as protest movements such as BC's "war in the woods" and New Brunswick's "war on the water" over fishing rights in Miramichi Bay, has been a significant transfer of formal and informal power over resource management and development on traditional territories to individual Ab-original communities (Hayter 2003; Fox 2006). We have also seen that empowerment has unmasked the pluralism of Aboriginal communities and their stances towards development. The great victory of these court battles and public activism has been to expand Aboriginal rights beyond the *Indian Act* and (however informally) beyond reserve lands to encompass traditional territories. In the process, however, fissures and divisions over directions for development have appeared. Some First Nations groups that strongly oppose salmon aquaculture express their discomfort with neighbours who have partnered with the industry. On the other side, communities involved in aquaculture sometimes return the criticism, again stressing the question of rights.

It's been very, very awkward, because many of us are relatives to that community [Klemtu] ... You know, I find it really awkward. When their community has to come through our community for medical services, we're always there to help them. When fish farming came into place, all of a sudden the invisible boundaries that we put in place many, many years ago – we shared those boundaries, it didn't matter if we went across each other's land ... Ever since fish farming was put into place, I feel that there has been a big wedge put among our communities, and it hurts, because our relationships are not the same anymore. We have to find a way to correct that. (Gary Housty, Hereditary Chief, Heiltsuk Nation, spoken at Bella Bella, BC)

I'd like to ask society and other [First] Nations: please leave us alone. We're doing everything good ... We're here to work for our people, to create employment and protect our resources. That's why we're here. I'm asking society to please leave us alone – other societies, other [First] Nations. We're doing the things we want to do in our territory. Let us do it. That's all I'm going to ask. Leave us alone. Let us carry on with what we're doing. (Archie Robinson, Hereditary Chief, Kitasoo-Xai'xais First Nation, spoken at Klemtu, BC)

The tension among First Nations communities further complicates the "labyrinth" facing the aquaculture industry at the local level. Returning to Schreiber and Newell's observation (2006a), neither support for nor opposition to aquaculture is clear-cut in First Nations groups or communities. While opposition groups and industry associations often frame Aboriginal stances in absolute terms, the local reality is that opposition to and support for aquaculture are complex. As we have seen through the words of First Nations leaders and community members themselves, local understandings of aquaculture draw on deep themes that reflect Aboriginal grievances on issues of rights and governance. Importantly, these themes are articulated in narratives for and against aquaculture development. If aquaculture is indeed here to stay in Canadian waters, this lesson needs to be learned by both governments and industry. It means that Aboriginal opposition to the industry is in some respects fundamental. That is, the entanglement of the industry with associations of colonial violence, threats to subsistence, and violation of rights (as well as risks to environmental and human health) will not be easily undone. At the same time, however, we have seen that industry efforts at partnership with Aboriginal communities have been symbolically and substantively successful in some cases. Although these partnerships remain controversial even within the communities that have chosen this path (Schreiber 2002), these cases demonstrate the value of direct negotiation between firms and individual communities as a means of navigating the controversy. Partnership does not mean that the explosive themes in the Aboriginal encounter with aquaculture are absent or subdued. Rather, we have seen that these are just differently articulated – in a way that highlights continued activism in the context of partnership.

This suggests that First Nations *support* for aquaculture development does not translate into *licence* for aquaculture development. This is an important distinction. Some supporters of aquaculture, most notably in government, discuss the need for aquaculture to earn "social licence" in order for the industry to counter controversy and grow in coastal regions of the country (Bastien 2004, 33-34). "Social licence" essentially refers to an acceptance of the industry as a legitimate user of the coast, and a trust that the industry is behaving responsibly and in the best interests of stakeholders and consumers (Gunningham et al. 2004). In a more utilitarian sense, it refers to the stocks of goodwill and trust – both in local settings and in a generalized sense – that an industry can draw on to act despite opposition and resistance (cf. Joyce 2000; Tollefson and Scott 2006). The conclusion that we draw, however, is that the notion of social licence is precarious when it comes to aquaculture development in First Nations communities and traditional territories. As we argued above, Aboriginal participation in aquaculture appears to be an assertion more than an endorsement. It is a furtherance of activism (on issues of livelihood, dependency, and rights) rather than its end. From the quotations provided, there is evidence of emerging trust within these partnerships. The notion of licence, which implies stability, permanence, and a unidirectional orientation in this trust, misses the activist nature of both Aboriginal protest and participation in aquaculture.

Aquaculture as Collision and Metaphor
Resource development in this country has recently become very complicated. Blessed with tremendous resource wealth, Canada experienced the rise of conservation and regulation in major resource sectors relatively late. Once entrenched, however, the authority of government and industry to dominate resource management and development has rarely been challenged. The predominant method of resource development – involving centralizing authority, exploiting environments to their maximum yield, and ignoring local and Aboriginal rights claims – began to change only in the late 1980s

and 1990s. Nor have these changes been wholly voluntary. Economic globalization is forcibly changing Canada's traditionally Fordist resource economy. Neoliberalism, whether embraced by choice or necessity, is changing the relationships among state, industry, and communities. Environmental degradation and the environmentalist movement are changing the cultural and political context of resource management. And the Aboriginal rights movement is meeting governments and industry head-on, forcing both a shift in overall approaches to development and a pluralization of how development occurs in local and regional spaces.

Into this steps aquaculture, the first genuinely new resource industry in Canada in a long while. This is an industry that is being directly shaped by these forces from the outset. Although more established sectors such as agriculture, forestry, mining, energy, and capture fisheries are confronting the same intersection of global and local pressures, aquaculture faces these without a tradition, with neither an established political economy nor a cultural presence. As discussed in Chapter 1, its late emergence created a *tabula rasa* with regard to aquaculture on the part of stakeholders, policy makers, and the general public. As an industry that does both familiar and unfamiliar things, it appeared as a "blank slate" that could be filled in many different ways. We argue that because aquaculture has arisen at this moment of significant intersecting global and local changes, it has become a locus for the expression of deeper conflicts over industrial development. For local people, aquaculture has become an important but fluid metaphor for both hopes and fears, dreams and nightmares. This explains much of the thematic consistency in local articulations of opposition to and support for aquaculture, but it also clearly perpetuates the controversy, as it charges the debate with a high level of emotion over the future of coastal life.

Part 2
Knowledge Battlefield

Although aquaculture occurs mainly in the periphery in Canada – in coastal communities and regions that are often far removed from urban centres and the levers of political power – the industry is embroiled in some of the most significant challenges facing the country at the turn of the century. As we saw in Part 1, these challenges include the extension of economic globalization, ongoing reforms to governance linked to the ascendancy of neoliberalism, the rise of global environmentalism, and the growing Aboriginal rights movement. Each of these forces leaves distinct marks on the aquaculture industry in Canada.

In Part 2, we argue that the aquaculture controversy is also entwined in the changing politics of knowledge. For most Canadians, aquaculture is learned rather than lived. This means that, outside of rural and coastal areas, the conflict over aquaculture is fundamentally a *knowledge conflict*. This conflict involves many actors who make or communicate often competing claims about aquaculture, including the news media, pro- and anti-aquaculture activist groups, firms, stakeholders, university scientists, and government agencies.

In the following three chapters, we look for patterns in claims making and examine the often bitter exchanges that have occurred between pro- and anti-aquaculture groups. At a deeper level, however, we argue that the persistence of the aquaculture controversy in Canada is linked to broader changes in the politics of knowledge, specifically in the way that knowledge claims are produced, communicated, and received by experts and the general public. Some of these changes are *structural*. For instance, we argue in Chapter 3 that the production of scientific knowledge has changed substantially over the past half-century in a manner that has given non-traditional actors (particularly activist groups) a strong scientific voice. Other changes are primarily *cultural*. For instance, local and traditional forms of knowledge – once thought to be irrelevant in the rational management of natural resources – are being reconsidered by politicians, experts, and the general public.

Together, these changes point to the pluralization of access to "authoritative" knowledge. In brief, this means that multiple groups are able to make strong claims about aquaculture that are capable of drawing the attention of media, activists, and the public. As we shall see, traditional authorities such as governments and industry are still learning how to cope with the consequences of this shift in the politics of knowledge.

3
Knowledge Battlefield: Science, Framing, and "Facts"

We are constantly told that we live in a "knowledge society," where knowledge and information are widely available to citizens and consumers, and that our increasing capacity to generate and manage knowledge will improve productivity, performance, and flexibility in both government and industry (Castells 2000). In short, the current era is one where knowledge and information are lionized and pursued as never before. As a society, we are placing a great deal of faith in knowledge to solve problems and improve our quality of life.

At the same time, however, many social scientists argue that we are living in an age of uncertainty (e.g., Beck 1992, 1999; Giddens 1990, 2000; Nowotny et al. 2001; Strydom 2008). As our knowledge of the natural world becomes more sophisticated and refined, the consequences of prior decisions are becoming more difficult to ignore. Decades of unrestrained industrial development and poor resource management have wreaked havoc on the environment and on human health in ways that we are only beginning to understand. According to Beck (1992, 60), we have now arrived at a point where "the side effects are taking centre stage" – where the dark side of development is beginning to loom large in the minds of policy makers and the general public. Importantly, much of this fear is based on things that we don't know, such as the long-term effects of pollution on the environment and the human body (Macnaghten and Urry 1998). These uncertainties are a significant motivator of environmentalist and consumer activism (Yearley 2005).

In this chapter, we argue that the tension between knowledge and uncertainty is strongly influencing the aquaculture controversy in Canada. Aquaculture has been the subject of extensive scientific investigation for over twenty years as part of a concerted effort to improve the environmental and economic performance of the industry. The exponential generation of knowledge about aquaculture has not defused the controversy, however; if anything, it has had the opposite effect. Conflicts over competing claims to authoritative knowledge about aquaculture are frequent and intense. To borrow a term from Slovic (1999), the aquaculture controversy in Canada can be understood as a "knowledge battlefield," where proponents and opponents of the industry struggle to assert the validity of their claims while delegitimizing those of their adversaries.

The aim of this chapter is to examine how and why this occurs. There are three parts to our analysis. First, we discuss how the aquaculture controversy is intertwined with the recent history of the natural sciences. Second, we examine the struggle between anti- and pro-aquaculture groups to frame or create narratives about aquaculture. Third, we look at how these groups mobilize claims about *facts* and *uncertainty* in attempts to influence public understanding of the industry.

Aquaculture and the "Pluralization" of Science

Science is the primary language of the aquaculture controversy in Canada. Proponents and opponents of the industry base much of their claims on scientific findings, referring emphatically to this or that study, this or that proof. These competing claims are notable in their incommensurability and finality: each side routinely claims to have firm scientific grounds for its positions that trump the claims of its opponents. In this section, we look at recent developments in the evolution of the natural sciences to begin assembling an explanation of how it is possible for multiple and incompatible science-based claims about aquaculture to gain and maintain credibility over the long term.

Science is one of the cultural pillars of modern Western societies. The eminent turn-of-the-century German sociologist Max Weber understood this significance, and famously argued that science is both a form of knowledge and a set of cultural values (Weber 1949 [1919]). These values, which include rationality, neutrality, and community (as in the open sharing of knowledge and methods), have long anchored Western understandings of science. This cultural view of science is that of a dispassionate, cumulative, and linear project:

> In science, each of us knows that what he has accomplished will be antiquated in ten, twenty, fifty years. That is the fate to which science is subjected; it is the very meaning of scientific work ... Every scientific "fulfilment" raises new questions; it asks to be surpassed and outdated ... Yet they will be surpassed scientifically – let that be repeated – for it is our common fate, and more, our common goal. We cannot work without hoping that others will advance further than we have. In principle, this progress goes on ad infinitum. (Weber 1949 [1919], 138)

Weber's characterization of "science as a vocation" has proven to be resilient. While many scientific breakthroughs and advances during the twentieth century were the product of secret and proprietary science (either military or corporate-sponsored on both sides of the global Iron Curtain), the ideology of science as a progressive and common project undertaken for the benefit of humanity has endured.[1] This can be seen in the works of Robert K. Merton, the most influential sociologist of science in the mid-twentieth century, who argued that scientific practice is bound by a normative code that is non-capitalist and apolitical. According to Merton, this code was based on the principles of communism (communal access to scientific knowledge and discovery), universalism, disinterestedness, and organized skepticism (Merton 1973 [1942]). Despite the deep involvement of science with political, military, and industrial projects throughout most of the twentieth century, both the general public and the scientific community held to a view of science as being somehow detached from these exigencies.

In recent years, however, assumptions about the uniformity, value-neutrality, and linear nature of science have been challenged. Social scientists are at least partly to blame for this. Beginning in the 1970s, the "sociology of science," associated chiefly with Merton, fell out of academic favour and was supplanted by a "sociology of scientific knowledge" (Collins 1983). This involved an important change in emphasis:

> The [older] sociology of science ... [aimed] to elucidate the set of normative and other institutional arrangements that enable science – the asking and answering of questions about Nature – to exist and function efficiently. A crucial feature of this program of inquiry is the assumption that the ultimate answers to the question are Nature's, mankind being only a mediator ... This program does not require sociological attention to the content of scientific answers ... The sociology of scientific knowledge, on the other hand, is concerned precisely with what comes to count as knowledge and how it comes to count. (Collins 1983, 266-67)

Pioneers of the sociology of scientific knowledge, including Barnes (1974), Collins (1974), Bloor (1976), Latour and Woolgar (1979), and Knorr Cetina (1982), began exploring how scientific knowledge is created and disseminated. This work has been controversial, as some commentators see these ideas as undermining the authority and legitimacy of scientists and scientific knowledge (see Sokal and Bricmont 1998; Hacking 1999).

Recent challenges to traditional understandings of science have also come from changes internal to the natural sciences. The views of science captured in the works of Weber and Merton were based on an assumption of exceptionalism. That is, the assertion of "science as a vocation" (Weber) or of "science as a normative structure" (Merton) assumed that the scientific community was for the most part uniform, capable of enforcing boundaries (membership coming only through extensive education and rigorous socialization), distinct from other fields and motives (such as military and industrial), and more committed to the values of the scientific community than to competing interests, such as politics and wealth. In other words, the traditional twentieth-century understanding of science viewed it as *exceptional* and *boundaried*. Although contemporary historians and sociologists of scientific knowledge now question whether this was ever actually the case, several recent developments have clearly served to "unbind" or pluralize science, making it both internally and institutionally more complex. We will first review these changes, and then discuss how they make protracted knowledge conflicts over issues such as aquaculture possible.

The Rise of Ecology as a "Subversive Science"

We argue that science has become unbound across the three key dimensions that are summarized in Figure 3.1. The first dimension is the rise to prominence, in the 1960s and 1970s, of ecology as a new scientific discipline that contains an implicit challenge to other areas of the natural and applied sciences. The basic idea behind the science of ecology – that organisms be studied in their environment and in relation to other living things – is an old notion that can be traced back to Aristotle (McIntosh 1985, 10). Ecology as a scientific discipline and distinct area of study did not emerge until the late nineteenth and early twentieth centuries, however, and remained a minor field in the natural sciences until the 1960s (Worster 1994; Hannigan 2006, 43).

Several scholars have argued that the emergence of ecology was a watershed in the history of the natural sciences (e.g., McIntosh 1985; Worster 1994; Forsyth 2003; Hannigan 2006), particularly because ecology's attention to context and interrelationships is at odds with the reductionist approach that has dominated modern scientific inquiry (Forsyth 2003, 5). Reductionism involves the simplification or breaking down of objects and

Figure 3.1

Dimensions of change in science and its institutional context

Change	Consequence
Rise of ecology and ecosystems science	Science-based critique of science
Institutional diffusion of science	Introduction of new scientific voices
Rise of "inclusive science"	Recognition of non-scientific expertise

problems to their simplest components in order to better understand the fundamentals of observed phenomena and thus improve the generalizability of findings (Gallagher and Appenzeller 1999; Suzuki and Dressel 1999, 45). Ecology's emphasis on systems and environments resists reductionism and tends towards a more holistic or complex view of the phenomena under examination (Ward and Dubos 1972; Lawton 1999).[2] The rise of ecology and holism is also notable because it belatedly brought human impacts into environmental science (Forsyth 2003, 5). Although most early ecologists focused exclusively on understanding ecosystems in and of themselves, this eventually led to concerns over how such ecosystems were being disrupted or changed. Beginning in the 1960s, human interference in the biophysical world became an overarching theme in ecological research and education (Worster 1994, 340).

According to Forsyth (2003, 4), this combination of resistance to reductionism and de facto critical stance towards human activities meant that ecology, at the time of its rise to prominence in the postwar period, had a tense and even "subversive" relationship with other fields in the natural and applied sciences (Sears 1964; Shepard and McKinly 1969). This new science posed a strong challenge to traditional narratives of technological and industrial progress, as well as to the role of scientists within these narratives. According to Worster (1994, 340):

After two centuries of preparation, ecology burst onto the scene during the 1960s. By then scientists of every sort were accustomed to appearing as society's benefactors. They were expected to show nations how to increase their power and citizens how to increase their wealth. But [ecology] took on a new role in a more nervous, anxiety-ridden time ... [based on] a grim hopefulness that ecological science would offer nothing less than a blueprint for planetary survival.

Today, ecology is no longer a "radical science," in the sense that many of its ideas and principles have been incorporated into the practices and training of scientists in other disciplines (cf. Williams 1990; Gray 2004). Nevertheless, the rise of ecology as a science and as an ideology has been central to the establishment of what Beck (1992, 156) has called "a science-based critique of science" that was not widely present or accepted prior to the 1960s. Simply put, ecology established "the consequences of science and technology" as a legitimate field of scientific study. Not surprisingly, this new field resonated with the nascent environmentalist movement that was developing its own critique of industrial development (Worster 1994). Ecology became a way for both experts and activists to talk against the excesses of the applied sciences in particular, using the tools, language, and authority of science itself (Hannigan 2006, 45).

The Institutional Diffusion of Science

The second change outlined in Figure 3.1 is the recent institutional diffusion of science. Overall, the nineteenth and twentieth centuries were a period of widespread professionalization and institutionalization in scientific inquiry. The natural sciences have always had strong institutional links, particularly to universities, private industry, and (more recently) governments. A remarkable number of early scientific discoveries, however, came from individuals, groups, and societies with few or no appreciable links to universities or other institutions (Bryson 2003). Prior to the twentieth century, much scientific practice could be described as artisanal, pursued both by accredited scientists with recognized academic posts and by a motley of passionate and often self-funded amateurs and hobbyists (many of them aristocrats) working out of their homes and personal laboratories (Nye 1996; Russell 2007).

The writings of Weber and Merton on the exceptionalism and boundaried nature of science are reflective of a long process of professionalization. Beginning in the nineteenth century, formal scientific credentials, although always important in garnering the respect of peers, were increasingly used to restrict access to patronage (and later to research grants), as well as to publication opportunities, laboratory resources, and membership in scientific associations (cf. Mendelsohn 1964; Reingold 1976). Increasingly, amateurs and the unaffiliated were actively excluded from the vocation as "professional scientists" sought to improve their standing in universities and attract stable government and private support (Mendelsohn 1977, 22).

The high-water mark for the institutionalization of science occurred during and after the Second World War. The war itself brought unprecedented investment into select areas of the natural sciences, particularly physics, but it was the postwar period in particular that saw a massive expansion of institutionalized science. Especially in North America, governments invested heavily in new and expanded universities as part of the demobilization of their armed forces. At the same time, science-based government agencies were established or greatly expanded, such as the National Aeronautics and Space Administration (NASA) in the United States (founded in 1958) and the National Research Council in Canada (founded during the First World War and greatly expanded in the 1940s and 1950s).

This institutional growth prompted Weinberg (1961) to famously suggest that the world had entered a new era of "Big Science," where scientific practices and discovery were becoming increasingly dependent on structured funding and other institutional supports. Similarly, Price (1963, 1) argued that the new scale of scientific investigation and its reliance on large institutions were entrenching new scientific norms:

> Because the science we have now so vastly exceeds all that has gone before, we have obviously entered a new age that has swept clear all but the basic

traditions of the old ... The large-scale character of modern science, new and shining and all-powerful, is so apparent that the happy term "Big Science" has been coined to describe it. Big Science is so new that many of us can remember its beginnings. Big Science is so large that many of us begin to worry about the sheer mass of the monster we have created. Big Science is so different from the former state of affairs that we can look back, perhaps nostalgically, at the Little Science that was once our way of life. (Price 1963, 2-3)

Recently, however, the "Little Science" that Price refers to is once again becoming a force in the natural sciences. More precisely, it appears that we are in the midst of a "re-diffusion" of science, where scientific knowledge is increasingly being generated, manipulated, and communicated by small, non-mainstream actors that are outside of, or only indirectly associated with, the main twentieth-century scientific institutions of academia, government, and private industry.

The institutional (re-)diffusion of science and expertise has several causes, the most obvious being the growth of science itself. Science education has grown tremendously since the middle of the twentieth century, as the number of students enrolled in Canadian universities in the natural sciences and engineering has increased from under 10,000 in 1950 to over 182,000 in 2005 (Wisenthal 2003; Statistics Canada 2005). With the increasing number of postsecondary graduates trained in science and engineering, it is not surprising that many of these find occupational homes outside the traditional institutions of academia, government, and industry.

A second reason for the recent institutional diffusion of science involves changes in scientific, information, and communication technologies. As discussed earlier, the rise to prominence of ecology as a scientific discipline literally opened up the world and all of its systems for scientific analysis. More recently, improvements in data-gathering technologies are enabling environmental science to be conducted more cheaply and further away from the centres of the scientific world. The costs of common environmental field tests have declined significantly over the past thirty years, at the same time that these tests have become more sensitive and sophisticated (cf. Chowdhury and Jakariya 1999; Che 2005). Moreover, new information and communication technologies are making collaborations and the pooling of data easier. In short, technological changes are enabling small players to participate in data generation and analysis at minimal cost. As we will discuss later, environmentalist groups in particular have taken advantage of new scientific resources (human and technological) to produce and communicate scientific claims about aquaculture that have successfully established alternative narratives to the "official science" of government and industry.

The Rise of "Inclusive Science"

The third dimension of change in Figure 3.1 is the increasing pressure on scientific practitioners and institutions to engage in a more "inclusive" science. In recent years, the presumed barriers between science and non-scientists have been challenged by citizen activism and new government directives for transparency in governance and decision making (Irwin 1995; Gunningham et al. 1998; Howlett 2000). Whereas decision making about issues such as pollution, resource management, waste disposal, and infra-structure development used to be done behind closed doors by politicians and experts, most democratic societies now engage in often lengthy processes of public consultations, hearings, and negotiations in order to legitimize decisions on contentious environmental or development projects (Fischhoff 1995; Leiss 2000; Nowotny et al. 2001).

Consultation and other public processes put pressure on scientists and scientific knowledge. First, the public airing of scientific claims has lessened overall public deference to "official" science and expertise (Irwin 1995; Ali 1997; Yearley 2000). Scientists and experts often use the language of prob-abilities when describing the rationale behind policy options or decisions. For example, experts will rarely assert that there is no risk associated with a given industrial activity, but will instead, in a manner consistent with sci-entific methods and professional norms, make statements about "minimal risk" or "an absence of evidence of risk" (cf. Irwin 1995). While these state-ments may be technically accurate, they often grate against the priorities of concerned members of public who want clear answers about safety and consequences (Slovic 1999; Fischer 2000, 134). This dissatisfaction can lead to skepticism and cynicism, particularly when alternative scientific claims and narratives are available (following the institutional rediffusion of sci-ence) from environmentalist groups or other non-mainstream actors. In addition, public deference to official science is being challenged by know-ledge society technologies such as the Internet. The broad availability of specialized information and opinion on the Internet means that members of the general public have the capacity to become sophisticated "lay experts" themselves on a range of contentious issues, and many are willing to use public consultation processes as means of questioning and challenging the official government and industry line (Ali 1999).

At the same time that environmental science is becoming more public and publicly debatable, the movement towards public consultation is also increasing pressure to incorporate different kinds of knowledge, particularly local and traditional knowledge, into formal scientific processes such as environmental assessments and risk management strategies (Young and Matthews 2007a). This is the antithesis of the exceptional and boundaried science described by Weber and Merton. The notion that scientific and lay knowledge may be integrated is new territory (Nadasdy 1999). Traditionally, it has been presumed that these are mutually alien, and even mutually unintelligible. While calls for better science communication have been made since the 1950s, these pleas generally assumed that scientists needed to do a better job of getting the public's attention, rather than the other way around (Irwin and Wynne 1996).

Research shows that the scientific community is divided on the desirability of incorporating non-expert claims into formal science-based decision making. While many scientists are open to the idea of public participation (Young and Matthews 2007a), others "have shown considerable and deep resistance to reconsidering their [opinions] of the public and ... [its] knowledge" (Wynne 1993, 321). What are truly lacking, however, are recognized ways of integrating scientific and public knowledge (Collins and Evans 2002). Without these mechanisms, participants must rely on mutual goodwill in order to muddle through the integration of different forms of knowledge (Petts 2008). Too often, it appears to stakeholders and other participants that experts are not taking consultation processes seriously – that they are listening impassively rather than engaging public claims (Kerr et al. 2007). Unfortunately, this failure to engage only tends to deepen public skepticism of official claims and to make people more receptive to alternative arguments and conclusions (Ali 1999; see also Rowe and Frewer 2000).

In summary, there is now sustained pressure to open up scientific processes – to make expert decision making more transparent, inclusive, and sensitive to non-scientific forms of expertise, particularly local and traditional knowledge. Although the mechanisms for doing this are underdeveloped, this pressure has increased public claims to and engagement in science-based dialogues and narratives (Irwin 1995; Fischer 2000; Kerr et al. 2007), further breaking down the exceptionalism and boundaried nature of science described by Weber and Merton. It is now much more possible for stakeholders and ordinary people to engage, oppose, and present alternatives to official science, and these claims have increasing political and procedural weight. Simply put, non-scientific forms of expertise are becoming more important in science-based decision making, and are further pluralizing the scientific landscape.

Aquaculture and the "Pluralization" of Science

These recent developments in the natural sciences make protracted knowledge conflicts such as the aquaculture controversy possible. The rise of ecology as a major scientific discipline, the institutional diffusion of science, and recent pressures to develop a more inclusive approach to expertise and science-based decision making have opened up room for multiple scientific stances and narratives about controversial industries such as aquaculture. As we will see later in this chapter when we look at specific knowledge conflicts between supporters and opponents of aquaculture, a great number of

groups are now generating and analyzing data about aquaculture's interactions with the environment and human health.

The plurality of actors wielding science means that the "degrees of freedom" for aquaculture research and claims making have increased dramatically. Environments are complicated objects of study, and ecological research
in particular involves making a series of choices about how to measure and
interpret data. These choices – about methodology, instruments, indicators,
geographic scale, and analytical emphasis – can dramatically shift data and
findings (cf. Southwood and Henderson 2000). Different actors will make
different choices, thus leading to a situation where incommensurable conclusions can be established and endure over the long term without one
disproving the other – at least in the minds of some "citizen-experts."

What does this mean at a deeper level for the aquaculture controversy?
As we shall see, it means that science is everywhere in this debate. Universities, governments, industry, and environmentalist groups are all involved
in the production and communication of conflicting scientific claims about
aquaculture. This leads to often bitter exchanges of claim and counterclaim,
science and counter-science. This reliance on science shows that the pluralization of science does not necessarily diminish its authority or credibility.
Because competing sciences are everywhere in this controversy, however,
the power of science *alone* to convince people of fact and truth is greatly
reduced. As Beck (1992, 156) has famously argued, the pluralization of science means that "science becomes more and more necessary, but at the same
time, less and less sufficient, for [establishing] a socially binding definition
of truth." In the remainder of this chapter, then, we will examine how narratives for and against aquaculture in Canada are created using both scientific
and moral arguments in attempts to outmanoeuvre competing claims on
the knowledge battlefield.

Creating Meaning: "Master Frames" on the Knowledge Battlefield
In this section and the next, we examine how pro- and anti-aquaculture
activists and groups labour to create, assert, and marry scientific facts with
cultural meanings. We will argue that this is done by pursuing both specific
knowledge (formal research) and broader frameworks of argument that cast
these specifics in particular ways (narratives and frames). Later in this chapter,
we will consider the "science wars" that have erupted on specific issues related to aquaculture (particularly over questions of the environment and
health). For the moment, however, our emphasis will be on the attempts of
pro- and anti-aquaculture groups to frame the aquaculture industry.

The term "framing" is used by sociologists to capture the deeper levels of
language and meaning that are mobilized by activists and institutions as
part of communication strategies (Gamson and Modigliani 1989; Benford
and Snow 2000). Framing is a strategic activity, where speakers use deliberate

language in attempts to link a subject with positive and/or negative cognitive associations. Framing typically involves making both specific and broad arguments. Specific arguments, which are occasionally termed "subframes," tend to cite facts that are not easily verifiable by the reader (this does not mean, however, that they are false or dishonest). These facts are then couched in "master frames" that contain broader arguments and connotations and that set the larger context for the specific facts and arguments presented.[3] For example, Reber and Berger (2005) discuss how oil industry lobby groups in North America have shifted framing strategies since the terrorist attacks of 11 September 2001. Prior to the attacks, the oil industry argued for expanded development using technical references to procedures such as "low impact drilling" (an environmental subframe) in combination with arguments about economic development, employment, and prosperity (master frame). Since 2001, however, the oil lobby has retained the environmental subframe but has shifted its master frame, now making broad references to terrorism, xenophobia, and nationalism ("energy independence") in an attempt to cast future oil development as a national security issue.

Many different groups are involved in attempts to frame the aquaculture industry either positively or negatively. Figure 3.2 outlines some of the most active groups. Although not exhaustive, it demonstrates the range of groups engaged in knowledge conflicts over Canadian aquaculture. As discussed in the Introduction, these groups are often formally and informally allied, which enables dominant messages to be communicated at local, provincial, national, and international scales.

We argue that *both* supporters and opponents of aquaculture employ three key master frames. Figure 3.3 identifies these as taint, risk, and injustice (oppositional frames), versus natural, modernization, and prosperity (supportive frames). As suggested by Figure 3.3, these master frames can be paired in the same manner as the local understandings of aquaculture discussed in Chapter 2.

Gamson and Modigliani (1989, 3) argue that five main devices are employed in creating master frames: "metaphors, exemplars (i.e., historical examples from which lessons are drawn), catchphrases, depictions, and visual images." These devices are ways of linking specific issues and information with broad themes. Indeed, Figure 3.3 shows that the master frames evoked by supporters and opponents of aquaculture are notably broad. Gamson and Modigliani (1989, 5) also argue that, for frames to be successful, they must have a strong cultural resonance: "Not all symbols are equally potent. Certain [frames] have a natural advantage because their ideas and language resonate with larger cultural themes. Resonances increase the appeal of a frame; they make it appear natural and familiar ... Some frames resonate with cultural narrations, that is, with the stories, myths, and folk tales that are part and parcel of one's cultural heritage."

Figure 3.2

Major nongovernmental organizations involved in the aquaculture controversy in Canada

Scale	Supporters of aquaculture	Opponents of aquaculture
Local or provincial	British Columbia Salmon Farmers Association British Columbia Shellfish Growers Association New Brunswick Salmon Growers Association Newfoundland Aquaculture Industry Association Prince Edward Island Aquaculture Association Aquaculture Association of Nova Scotia Aboriginal Aquaculture Association Positive Aquaculture Awareness First Dollar Alliance	Coastal Alliance for Aquaculture Reform David Suzuki Foundation Friends of Clayoquot Sound Georgia Strait Alliance Living Oceans Society Raincoast Conservation Society T. Buck Suzuki Foundation Watershed Watch Salmon Society Union of British Columbia Indian Chiefs United Fishermen and Allied Workers' Union Atlantic Salmon Federation
National	Canadian Aquaculture Industry Alliance Aquaculture Association of Canada	Canadian Centre for Policy Alternatives
International	Salmon of the Americas Global Aquaculture Alliance	Pure Salmon Campaign Greenpeace World Wildlife Fund Ecotrust

---

In the following subsections, we discuss the master frames of pro- and anti-aquaculture groups in light of Gamson and Modigliani's observation. At the same time, we argue that, at the current moment of heightened consciousness of risk and environment, as well as skepticism of industrial models of development, it is not immediately clear which master frames have the "natural advantage" assumed by these authors. At this moment in time, ours is a society that is both economically and environmentally sensitive. We therefore argue that both the oppositional frames (taint, risk, and injustice) and the supportive frames (natural, modernization, and prosperity) have the potential to resonate culturally. Using Canadian media coverage of aquaculture as an approximate measure, however, we find that oppositional frames are on the whole outpacing those advanced by pro-aquaculture activists and groups.

Figure 3.3

Master frames employed by nongovernmental groups in opposition to and in support of aquaculture

Opposition	Support
Taint	Natural
Risk	Modernization
Injustice	Prosperity

Taint versus Natural

We begin our discussion with the first pairing given in Figure 3.3: the master frames of "taint" and "natural." The notion of taint or impurity is culturally powerful. The celebrated anthropologist Mary Douglas demonstrated that the dialectic of purity and impurity or taint is one of the most fundamental conceptual divisions in human thinking – to the point that it directly impacts the organization of human groups (Douglas 1966). In some societies, this division is very evident, as in the South Asian caste system, where certain people are understood as sources or embodiments of impurity (as, literally, untouchable), whereas in other societies it is less discernible but nonetheless present. Douglas' main conclusion is that the idea of taint is used as a mechanism for drawing boundaries, between right and wrong, between self and other, between those included and those excluded from groups (114). Things that are unfamiliar, unusual, or ambiguous often become culturally defined as wrong, outside, tainted.

According to Douglas, the cross-cultural significance of taint signals its importance. Even in complex, secular, and science-driven societies, the ideas of purity and taint evoke strong emotions. This is particularly true with respect to the environment (cf. Douglas and Wildavsky 1982; Cronon 1995). Prominent environmental theorists such as Beck (1992), Wynne (1996), and Irwin (2001) suggest that one of the core drivers of contemporary environmental anxiety and activism is a fear of environmental perversion – specifically the fear that after decades of environmental pollution, the fundamentals of nature are being altered. Tainted nature, in the form of acid rain, ozone depletion, or nuclear radiation, is a perverted nature, taking health rather than giving it. These examples also illustrate the easy combination of science and purity/taint narratives. The scientific narratives surrounding acid rain and UV rays have a highly technical side, but they also reflect the ongoing salience of these broader (and older) themes (cf. Ungar 2000).

Opponents of aquaculture make common use of the taint master frame. Several research reports and other activist publications make explicit references to "unnatural" aquaculture methods and products, including the David Suzuki Foundation's report *Super-Unnatural* (Volpe 2001) and the US-based

Environmental Working Group publication *Factory Methods, Unnatural Results* (Environmental Working Group 2003). Anti-aquaculture activists have also developed slogans (or Gamson and Modigliani's "catchphrases") that emphasize the taint/unnatural theme. The most prominent of these is the Coastal Alliance for Aquaculture Reform's (CAAR) name for its anti-aquaculture campaign: "Farmed and Dangerous." Others include "Wild Salmon Don't Do Drugs" (the origin of this slogan is unclear, although it is currently reproduced by several Canadian and American organizations on T-shirts and bumper stickers), and "Farmed Salmon Dyed for You," which refers to the use of colourants in feed that are necessary to give aquacultured salmon its familiar pink hue. Advocacy materials such as pamphlets also stress these themes. For example, the following text is taken from a flyer distributed by CAAR (n.d.a):

> Farmed Salmon are fed antibiotics,
> colourants and pesticides.
> Bon Appetit.

> Salmon raised on farms are very different from wild salmon. For starters, they're raised in floating feedlots that pollute the ocean. They're fed chemical additives to make their flesh pink like wild salmon. Antibiotics and pesticides are used to control disease outbreaks on the farms. If that's not bad enough, farmed salmon contain disturbing levels of PCBs.

> Despite human health and environmental concerns, many restaurants and stores are still willing to sell farmed salmon to you – including some health and natural food stores you've come to trust. And that's enough to make anyone lose their appetite ... Go and tell the managers of your favourite grocery store and restaurants that you refuse to buy farmed salmon and ask them to refuse, too.

Figure 3.4 is a reproduction of a document posted on the CAAR website.[4] Viewers are encouraged to download the image and print it on a series of sticker address labels that can then be clandestinely affixed to aquaculture products in grocery stores. Again, the theme of taint is evident in the references to pollution, chemical contamination, and the use of pharmaceuticals on aquaculture products.

As suggested in Figure 3.3, supporters of aquaculture compete with the taint master frame with one that casts aquaculture as natural, and therefore healthy and environmentally benign. The "natural" master frame plays down any differences between aquaculture products and their wild counterparts. Thus, whereas opposition groups often frame farmed salmon as a

Figure 3.4

Reproduction of a document on the Coastal Alliance for Aquaculture Reform website (http://www.farmedanddangerous.org/)

perversion of natural processes, supportive groups frame them as equal, if not identical, to the more familiar commercially caught product. The following quotation from the British Columbia Salmon Farmers Association (BCSFA) illustrates this frame:

Today, many people are looking for healthy choices in their diet and life-styles. Fish are an important part of a well-balanced diet, and the Omega-3 fatty acids in salmon mean this product may provide some particularly compelling health benefits. And thanks to the salmon farming industry, fresh and high-quality salmon is now available year-round. Farm salmon are fresh, healthy products, rich in Omega-3 fatty acids, which can keep our bodies and minds healthy. (BCSFA 2005b)

One of the most controversial elements of the natural master frame in-volves the efforts of several aquaculture firms to classify their products as organic. The term "organic" is symbolically weighty, with connotations of a closeness to nature, non-interference in natural systems and processes, and the authenticity of small-scale or family agriculture (cf. Trewavas 2001; Mansfield 2004). For an industry that is routinely accused of marring the natural environment, the organic label is a potential godsend. The BCSFA argues: "Should salmon, farmed in the wild, and tracked from egg to plate, be eligible for organic certification when they are fed diets containing fish meal and grown in low density pens? We say, why not? Fishmeal is fed to organically certified land animals and net pens are no more than aquatic fences designed to contain the livestock" (BCSFA 2005c).

Organic certification of aquaculture products has proceeded in Europe, but not yet in North America. In 2003, a group called the Pacific Organic Seafood Association (POSA), whose members include some of the largest aquaculture firms that were then operating in Canada, as well as several smaller companies, made a submission to the Certified Organic Associations of British Columbia (COABC).[5] While the COABC application has since been abandoned, POSA has aligned itself with the standards of the European Union–based International Federation of Organic Agriculture Movements (IFOAM) as a means of claiming organic status.

The organic label is being strongly opposed by environmentalist groups, which view it as a means of confusing consumers and "greenwashing" the industry (e.g., CAAR 2005). The notion that aquaculture is claiming the organic moniker has also angered advocates of traditional fisheries. As a husbandry activity, aquaculture is in principle eligible for organic certifica-tion, whereas wild fisheries are not. Research conducted by Mansfield (2004) captures the incredulity of many fishers at this argument. To quote an Al-askan fisheries advocate: "How organic can you be? Wild fish roaming the oceans for food then spawning in their natal streams. No food source pro-vided by man, and no interference by man" (Mansfield 2004, 221). Organic trade associations have resisted giving organic labels to wild fisheries, how-ever, on the logic that "organic is not synonymous with natural" (Mansfield 2004, 221). According to one organic trade association: "The claim by the proponents of the certification of [wild] aquatic animals that their animals are organic simply because they are wild does not reflect the management required to produce a certified organic crop. The key to the use of the certified organic label on crops or livestock is the word 'management' ... In order to manage there must be control" (Mansfield 2004, 224).

Risk versus Modernization

The second master frame engaged by opposition groups is that of risk. Risk is a complicated notion that has recently begun to receive significant attention from social scientists (e.g., Beck 1992; Lupton 1999; Ericson and Doyle 2003). As discussed later in this chapter, the collision of competing and often highly technical understandings of risk plays a significant role in the knowledge battlefield over aquaculture in Canada. For the moment, however, we will briefly consider how risk is engaged as a broad master frame designed to achieve the cultural resonance described by Gamson and Modigliani (1989). Essentially, risk as a master frame is closely linked with questions of time (Adam 1998; Macnaghten and Urry 1998). According to Barbara Adam (1998), risks are most frightening to people when they are perceived to be both unpredictable and irreversible. Unpredictability and irreversibility form a powerful combination, suggesting that permanent change – to ecosystems, wild species, or human health – can happen at any moment. According to Macnaghten and Urry (1998), the combination of unpredictability and irreversibility has a particular cultural resonance because it reflects the two dominant understandings of time in a postmodern world. For these authors, the postmodern world is characterized by twin preoccupations with "instantaneous time" (exemplified by instant access to information and communications) and "glacial time" (exemplified by heightened awareness of long time scales surrounding environmental issues such as climate change). Macnaghten and Urry argue that the most contentious environmental controversies are those that encompass both instantaneous and glacial times. Nuclear radiation is the archetype, where "in each [random] instant, a nuclear reaction can unleash huge amounts of ... exceptionally destructive force; and that the effects of such forms of power will remain for thousands of years and almost certainly change the evolutionary development of the human species" (1998, 107).

Although the potential risk and damage posed by aquaculture is at a different scale from nuclear radiation, the dynamic of unpredictability and irreversibility is still potentially powerful in generating public opposition to the industry. The time-risk master frame is used in terms of both environmental and human health issues. With respect to the environment, the ultimate time-risk narrative is that of extinction of wild species. The spectre of extinction hovers over the Canadian aquaculture industry, as wild populations of its main species – Atlantic salmon – are severely threatened or extinguished in both North America and Europe. Although the main causes of these declines are overfishing and habitat destruction (Hindar 2003, 49), opposition groups often draw implicit links to the aquaculture industry. For instance, the Salmon Farm Protest Group (based in Scotland) argues that "what we are playing with here is a wild species that has existed on the planet since the end of the last ice age. We're pushing that towards extinction, and we're going to replace it with a totally artificial species" (Rabinovich 2007). The US-based Pure Salmon Campaign (n.d.) argues that "escaped fish transmit diseases and parasites to wild salmon, and threaten to establish viable colonies that could not only push wild salmon to extinction, but also related species such as steelhead and sea trout." The language applied to health effects is very similar. In the anti-aquaculture volume *A Stain upon the Sea,* activist Don Staniford writes (2004, 197-98):

> To avoid a "Silent Spring" of the sea we must curb chemical use in the entire sea cage fish farming sector now ... Over 40 years since Silent Spring was published, the chemical that Carson exposed, DDT, is still being found in farmed salmon ... It is a dreadful prospect to think what we will be finding in 2044 ... As Rachel Carson said on CBS television shortly before her death in 1964: man is a part of nature, and his war against nature is inevitably a war against himself.

Returning to Figure 3.3, we argue that supportive groups employ a "modernization" master frame to counter opponents' narratives of risk. Importantly, the modernization master frame also has a strong temporal element, but instead of having time represent the invisible, unpredictable, and irreversible, the modernization narrative emphasizes progress and evolution. As with most master frames, the modernization frame involves both general and specific arguments. At its most general, this frame attempts to link aquaculture with human history and progress. In this depiction, aquaculture is a natural extension of human ingenuity and stewardship over the natural environment, as can be seen in the following quotation from publicity materials of the New Brunswick Salmon Growers Association (NBSGA 2007):

> With easy access to oceans once believed to hold an endless supply of wild fish, few North Americans ever considered that anyone would need to "farm" fish. Yet people throughout the world have a long history of fish farming. Ancient Egyptians, Romans, Mayans, Aztecs, Hawaiians, Chinese, Japanese and many other sophisticated civilizations farmed finfish and shellfish and aquatic plants. As each of these civilizations grew, local populations of wild animals were exploited until hunting and fishing could no longer supply

the population's food requirements. Only through the domestication and farming of plants and animals could these civilizations grow and flourish.

Notably similar text is presented in the 2004 report of the government of Canada's Commissioner for Aquaculture Development: "Aquaculture, seen by some as the last step in mankind's 10,000-year journey from hunter-gatherer to food producer, has a lot in common with agriculture. There is the same delicate balance of living in harmony with nature while trying to guard against its capricious whims ... There is a seasonal rhythm, with no end to the work to be done" (Cayo 1993, quoted in Bastien 2004, ii).

The more specific modernization narrative is that the aquaculture industry is consistently modernizing or improving its performance on environmental issues. Individual firms, industry associations, and supportive non-governmental organizations frequently release pamphlets and other publications that emphasize technological advances and innovations (e.g., Mitchell et al. 2003; Neshevich 2005). As major sponsors of aquaculture research and development, provincial and federal governments also promote the modernization frame (e.g., Bastien 2004; National Research Council 2005). The government and industry line stresses the market case for modernization, arguing that it is in "the best interests of [aquaculture firms] to provide safe, quality products for their clients" (Bastien 2004, 41), and that "it is in a salmon farmer's interest to operate in a sustainable fashion, especially in these ultracompetitive times" (Mitchell et al. 2003).

Injustice versus Prosperity

The final set of competing master frames presented in Figure 3.3 involves injustice and prosperity. We will consider these only briefly, as they are the most straightforward and, in a sense, uncomplicated of the major frames advanced in the aquaculture controversy. The injustice frame is a cornerstone of many popular social movements, from labour to women's rights to environment (cf. Carroll and Ratner 1996; Benford and Snow 2000). In the context of the aquaculture controversy in Canada, this frame highlights the environmental and cultural consequences of aquaculture development for vulnerable, local populations, particularly Aboriginal peoples:

For the past 10,000 years, coastal First Nations of British Columbia have relied on the sea to provide for their people. Supported by plentiful marine resources, First Nations communities developed vibrant cultures rich in song, dance and art works that are acclaimed worldwide. Abundant runs of wild salmon, together with the harvest of other seafood, fed entire villages. Village beaches still gleam white with sun-bleached shells, reminiscent of bountiful harvests ... All this is in jeopardy. The establishment of salmon farms on the BC coast has had a disastrous effect on First Nations' ways of life. (CAAR n.d.b)

The injustice frame differs from the other two oppositional narratives in that it rests primarily on a moral argument. Whereas the frames of taint and risk make a personal appeal to consumers, the injustice frame asks observers to imagine the consequences of industrial development to others. As discussed in Chapter 2, notions of justice and injustice are crucial to *local* understandings of aquaculture. As we shall see, however, the injustice frame is the *least* effective of the oppositional narratives, at least in terms of breaking through the print media.

In contrast, the prosperity frame is the most successful of the pro-aquaculture narratives. According to this depiction, aquaculture confers gain on suffering regions and groups. As such, the prosperity frame is both a moral and instrumental narrative. For instance, in the pamphlet "Salmon Farming, the Key to Coastal Revitalization," the BCSFA states: "Clean air, beautiful scenery and an active lifestyle ... salmon farming provides rewarding career opportunities that make it possible to live and work in coastal British Columbia" (BCSFA n.d.). The prosperity narrative extends beyond the moral association with depressed coastal regions and marginalized populations. The Canadian Aquaculture Industry Alliance (CAIA) argues that with proper government attention, "within the next 15 years, Canada's aquaculture output could reach $2.8 billion annually ... and provide sustainable, year-round employment for more than 47,000 people" (2005b, 2). The Commissioner for Aquaculture Development – the most senior federal civil servant tasked with promoting the expansion of the industry – opened his 2004 report *Recommendations for Change* (Bastien 2004) with the following quotation from Nobel laureate in Economics Peter Drucker: "Aquaculture, not the Internet, represents the most promising investment opportunity of the 21st century."

Successful and Unsuccessful Frames

Framing is about strategic communication. As such, it involves creating narratives, weaving together specific facts (that are often difficult for readers to verify) with much more general claims, in attempts to create cognitive associations. But while framing is about creating a deeper context, it is based on simplification. At its root, it is about creating linkages and associations that leave out complexities and competing understandings. As we have attempted to show in this section, master frames invoked in opposition to and in support of aquaculture are in direct tension and indirect conversation. In the broadest terms, the oppositional master frames of taint, risk, and injustice cast the industry as something that is unnatural – as a contravention of the way that things ought to be, in both an environmental and a social sense. Against this, the supportive master frames of natural, modernization, and prosperity represent attempts to frame the industry as something comfortable and familiar, within the linear story of human and technological progress.

It is difficult to judge the effectiveness of these narratives in swaying public opinion. To begin with, it is often difficult to situate these arguments in the wider public discourse about aquaculture. The first reason for this is that, although we have argued that there are three main distinctive frames being advanced for and against aquaculture, these narratives are often intertwined. As Gamson and Modigliani (1989, 2-3) argue: "Public discourse is carried on in many different forums. Rather than a single public discourse, it is more useful to think of a set of discourses that interact in complex ways ... A [single position] typically implies a range of positions, rather than any single one."

For example, taint and risk narratives are not always presented separately or distinctly. Although we have argued that they have different cultural resonances, these arguments often blend into one another. Furthermore, we ought to dispel any undue assumptions about the coordination and consistency of the messages being communicated by pro- and anti-aquaculture groups. As demonstrated in Figure 3.2, literally dozens of groups are actively engaged in communications on either side of the issue. While these groups are often networked or otherwise connected to one another (which enables their messages to reach broader and more diverse audiences across geographic scales), these networks do not produce singular sets of messages. There is clearly coordination, but many comments, publications, and other forms of activism are often produced ad hoc by individual groups acting on their own.

The second reason why it is difficult to judge the effectiveness of the competing frames in the public discourse is the unreliability of survey data on controversial issues such as aquaculture. Some researchers rely heavily on public opinion polling to judge the effectiveness of framing and communications strategies (e.g., Kinder and Sanders 1990; Iyengar and Simon 1993). There are serious pitfalls, however, when one tries to measure *understanding* and *meaning* with survey data. Gamson and Modigliani encountered this in their seminal research into the framing of nuclear power in the United States:

> Public thinking on nuclear power is very difficult to test directly with existing survey data. Consider the dilemma that a typical survey question presents to respondents that hold a devil's bargain [view of the nuclear industry]: "In general, do you favour or oppose the building of more nuclear plants in the United States?" How does one respond if one believes that nuclear power is a necessary fact of life but that sooner or later there will be an enormous price to pay? Does one answer favour, oppose, or not sure? (1989, 34)

The aquaculture controversy itself demonstrates these difficulties. In April 2007, CAAR released the findings of a public opinion survey that they had commissioned from the firm StratCom Strategic Communications. The survey of 606 citizens of British Columbia found extensive opposition to salmon aquaculture (StratCom 2007). The questions posed, however, were clearly intended to produce responses that cast aquaculture in a negative light. For example, several questions were prefaced with the following: "As you may know there has been public debate about the farming of salmon here in BC. In particular, many critics have pointed out that current technologies which are supported by the government do not separate the wild salmon stocks from the farmed salmon stocks, so the two populations mingle and that is hurting the wild salmon." Moreover, the questions themselves were designed in a manner that reflects the risk frame: "Do you agree or disagree that the government should stop supporting open net cage salmon farming and instead start investing in closed containment technology that separates farmed salmon from wild salmon and protects the marine environment?" Predictably, this question yielded nearly 81 percent agreement (StratCom 2007). Pro-aquaculture groups are involved in the same deliberate skewing of survey research. In May 2007, a month after the CAAR release, the BCSFA distributed a media release highlighting a poll conducted by Ipsos Reid but sponsored by the BCSFA. In this case, the question "Do you support or oppose the development of a fish-farming sector in BC that will not negatively impact the environment?" yielded 65 percent support (BCSFA personal communication 2007). This question conflates two distinct issues – the expansion of aquaculture and environmental integrity – which violates the methodological rules for reliable social research (Gray and Guppy 1999, 100). Because of this, neither CAAR nor the BCSFA results are particularly meaningful.

Setting aside the question of public opinion, we will rely on a content analysis of print media coverage in order to gauge the effectiveness of the master frames advanced by pro- and anti-aquaculture groups. This method has been profitably used in other studies as a means of investigating *public discourse* rather than public opinion (e.g., Gamson and Modigliani 1989; Powell and Leiss 1997; Rusciano 2003). We agree with Gamson and Modigliani (1989, 3), who argue that "mass media are not the only forums for public discourse, but, since they constantly make available [frames] and meanings and are highly accessible in media-saturated societies ... their content can be used as the most important indicator of the culture surrounding a general issue."

Our media study will be the primary focus of Chapter 5, where we will examine findings in detail. For now, suffice it to say that the study involved the content coding of media items dealing with aquaculture (news articles, editorials, letters to the editor, and features) appearing in seven major

Canadian daily newspapers from 1987 to 2006 (a total of 1,558 items).[6] Examining the presence of master frames in these items requires some interpretation. While keyword searches are sometimes used in content analysis, they are inappropriate in our case. For example, the extent of the taint master frame could not be uncovered through keyword use alone (the word "taint" rarely appears in any item). Likewise, use of the word "risk" fails to distinguish between type of risk implicated (environmental risk and economic risk are often mobilized against each other in environmental controversies; cf. Young 2004).

Indicators are therefore used instead of keywords, and these are summarized in Table 3.1. For instance, we have taken "mention of health risk or harm from aquaculture" as an indicator of the taint master frame, given that the health-risk argument is based on the idea that aquaculture products are contaminated with toxins and additives. Similarly, we equate "mention of health benefits of aquaculture" with the natural frame, as the health-benefit argument nearly always stresses the similarities between wild and aqua-cultured product (especially with regard to the benefits of omega-3 fatty acids). "Mention of environmental risk or harm" is taken as an indicator of the risk frame. "Mention of ecological modernization" (industry efforts to reduce environmental impacts, and/or aquaculture as an improvement on traditional capture fisheries) is taken as an indicator of the modernization frame. Lastly, "mention of social or cultural risk or harm" (such as harm to local or Aboriginal rights and practices) represents the injustice frame, whereas "mention of economic benefits" is considered to be in line with the prosperity frame.

Table 3.1 also shows the number of media items that mention each indicator. Keeping in mind that these are imperfect measures, we nevertheless see evidence that opposition groups have had more success in creating meaning and finding cultural resonance on two of the three pairings of competing frames. Both the taint and risk frames outweigh their respective competing

Table 3.1

Media mentions of master frame indicators, 1987-2006

Indicator (frame)	Number of items with mentions	
Mention of health risk or harm (taint)	153	(9.8%)
Mention of health benefit (natural)	60	(3.8%)
Mention of ecological risk or harm (risk)	1,015	(65.0%)
Mention of ecological modernization (modernization)	64	(4.1%)
Mention of social or cultural risk or harm (injustice)	117	(7.5%)
Mention of economic benefit (prosperity)	626	(40.1%)

frames, with the ecological risk narrative in particular dwarfing the modernization depiction (mentions in 1,015 items, compared with 64). Only the prosperity frame, appearing in 626 items (roughly 40 percent of the total), outpaces its competitor, the injustice frame. These data do not tell us in a definitive way that anti-aquaculture activists are winning the struggle to frame the aquaculture industry in a negative light. They do suggest, however, that oppositional narratives about aquaculture have found purchase in the public sphere and pose a significant challenge to government- and industry-sponsored claims about the desirability of aquaculture development.

Science Wars

In this final section, we turn our attention to how pro- and anti-aquaculture groups *directly* challenge one another's claims about aquaculture science and "fact." The pluralization of science has made it possible for competing scientific narratives about aquaculture to coexist and compete over the long term. We argue, however, that the "science wars" over aquaculture are as much about *communication strategies* as they are about science. Looking at several cases of direct conflict over aquaculture science, we argue that opposition groups have been very successful in leading the *public* side of the scientific debate over aquaculture, forcing the industry and its supporters into a reactive stance – fighting to discredit anti-aquaculture claims rather than taking the lead in the public debate over aquaculture "fact."

Figure 3.5 outlines current scientific and expert disagreements over the practices and effects of aquaculture in Canada. As mentioned in the Introduction, the early days of the aquaculture industry in Canada were characterized by both a lack of controversy and a lack of specific knowledge about aquaculture. Operators came from varied backgrounds, and most learned about aquaculture procedures from books, pamphlets, and trial and error (Keller and Leslie 1996, 13). Scientific knowledge about aquaculture was borrowed, either from established activities such as hatchery-husbandry or from early research performed in more established aquaculture regions such as Norway. In the mid-1980s, however, the aquaculture industry in Canada encountered some serious problems stemming from a lack of good aquaculture science, including disease outbreaks, environmental events (such as algal blooms that killed fish by the tens of thousands), and record escapes due to poor engineering of net-pens.

Following the losses of the mid-1980s, governments, industry, and educational institutions began investing heavily in Canadian-based aquaculture science and expertise. The lead research agency in government has been the federal Department of Fisheries and Oceans (DFO); its provincial counterparts include New Brunswick's Ministry of Agriculture and Aquaculture, and British Columbia's Ministry of Agriculture, Fisheries and Food (Keller and Leslie 1996, 3). In addition to in-house research, governments have also sponsored

Figure 3.5

Current scientific and expert disagreements over aquaculture in Canada

Environment
- Possible spread of disease and/or parasites from aquacultured fish to wild fish
- Possible genetic pollution of wild fish populations due to escapes
- Possible displacement of wild fish populations due to escapes
- Effects of pollution from aquaculture operations on the marine environment
- Effects of pharmaceuticals use on marine life
- Effects of artificial lighting of aquaculture sites on marine life

Human health
- Particular harms or benefits from consumption of aquaculture products
- Net health effects (positive or negative) of consumption of aquaculture products

Rural economy
- Particular harms or benefits of aquaculture to specific industries and firms
- Net economic effects (positive or negative) of aquaculture on local economies

Rights
- Legal status of aquaculture claims to marine spaces
- Effects of aquaculture claims on Aboriginal rights

Sources: Gardner and Peterson 2003; McDaniels et al. 2006; Young and Matthews 2007a.

work in universities and private industry (see Chapter 7 for a full discussion). For instance, from 1999 to 2006, the government of Canada sponsored a large research initiative into aquaculture called AquaNet, which received over $10 million in total funding from Industry Canada, the Natural Sciences and Engineering Research Council, and the Social Sciences and Humanities Research Council. AquaNet sponsored large volumes of research, including the work presented in this book. In 2006, its membership totalled 176 students and researchers from Canadian universities, government agencies, and private firms.

Environmentalist groups and other anti-aquaculture organizations began to produce and sponsor alternative scientific projects in the 1990s. These projects have drawn on various disciplines of the natural sciences, particularly ecology, toxicology, and population biology. Occasionally, this research has been conducted by activist groups themselves, although most often funding is provided to third parties, including consultants and respected university-based scholars. Importantly, however, activist groups are nearly always at the centre of the communications strategies for moving these claims into the public and policy domains.

In the following subsections, we examine several instances of direct public conflict over aquaculture science and facts. We find that although each case deals with different facts and arguments, they follow broadly similar patterns. Consistently, anti-aquaculture groups have demonstrated the ability to dominate public discourses surrounding questions of science and fact, while supporters of aquaculture have found themselves following rather than leading public discussions of their own industry.

Net Loss–Net Gain

The first case that we consider is one of the earliest direct conflicts over the public communication of aquaculture science and fact. This conflict began with a 195-page report entitled *Net Loss: The Salmon Netcage Industry in British Columbia* (Ellis 1996). The report was commissioned by the environmentalist group the David Suzuki Foundation, and was authored by David Ellis, a professional planner and former commercial fisher. The Ellis report was not a bolt from the blue. Environmental concerns about aquaculture production had been voiced by industry observers and some DFO scientists since the mid-1980s (Keller and Leslie 1996, 31). A brief moratorium on the expansion of salmon aquaculture in British Columbia had been declared in 1986, and then lifted in 1987 following a review of evidence. In fact, the Ellis report was prepared as part of the David Suzuki Foundation's submission to a process called the Salmon Aquaculture Review (SAR), which was conducted in British Columbia following a second moratorium on new salmon aquaculture sites initiated in 1995. The SAR was not intended as an evaluation of the aquaculture industry per se, but rather as a means "to conduct a review of the adequacy of current methods and processes used by the [BC Ministry of Environment, Lands and Parks, and the BC Ministry of Agriculture, Fisheries and Food] in regulating and managing salmon aquaculture operations in British Columbia" (SAR 1997a, 1).

Net Loss was clearly intended to reach a wider audience than the panel of experts involved in the SAR, however. The David Suzuki Foundation published copies that it continues to distribute at minimal cost. The report itself is structured as a catalogue of objections to the industry, including chapters on the predominance of multinational firms in salmon farming, pollution issues, the use of pharmaceuticals, and disease transfer to wild fish. Numerous illustrations, tables, and charts summarize the arguments in easily understood graphical form.

Industry supporters did not react kindly to *Net Loss*. Immediately after its release, the BCSFA commissioned consultant Alan Kenney to formulate a response to the Ellis report. The result was a 61-page publication predictably entitled *Net Gain* (Kenney 1997). *Net Gain* is an unequivocal attack on *Net Loss*. It opens thus:

The Ellis report is a fundamentally flawed document that lacks credible academic rigor. The report presents a one-sided view of complex issues based on selective and inadequate research. The report also neglects essential information necessary for a full understanding of the issues. The BC salmon farming industry is therefore presented in an unrealistic and false manner. (Kenney 1997, 1)

In a pattern that would become familiar, *Net Gain* then proceeds with a point-by-point rebuttal of the Ellis report. Its chapters shadow those of *Net Loss*, and each chapter is internally organized to mirror the points made by Ellis. As an illustration of the point/counterpoint exchange, consider the following two quotations, the first from *Net Loss* and the second from *Net Gain*, both of which appear under the heading "The Issue of Fish Diseases":

Except at times during the final river migration and spawning period, wild fish do not encounter the high [population] densities found at netcage sites. This is the main reason that all diseases endemic to salmon find easier expression in netcage fish. The crowded netcage environment is stressful to salmon, and the higher the stress level, the higher the incidence of disease. The average densities now found in salmon netcage operations in B.C. are 5 kg/m³ for Chinook and 8-10 kg/m³ for Atlantic salmon. (Ellis 1996, 121)

Ellis begins this section with an unreferenced statement ... that is indicative of the bias and misrepresentation perpetrated in the report. On p. 121, Ellis states in reference to rearing densities in netcages for farmed salmon that: "This is the main reason that all diseases endemic to salmon find easier expression in netcage salmon." This assertion is not supported by scientific research. (Kenney 1997, 38)

The *Net Loss/Net Gain* exchange tells us several things about conflicts over aquaculture "facts" as well as the success of anti-aquaculture groups in leading this debate. To begin, it is significant that this particular conflict began with the publication of an anti-aquaculture document. This means that opponents of the industry were able to define the issues and knowledge claims that would need to be rebutted by the industry. To be clear, aquaculture, particularly salmon farming, was controversial before the publication of *Net Loss* (cf. Keller and Leslie 1996; Bailey et al. 1996). But according to extensive research by William Leiss and colleagues (Leiss and Chociolko 1994; Powell and Leiss 1997; Leiss 2001), the worst strategy for "risk issue management" is to allow opponents to define the issues at play (Powell and Leiss 1997, 217). Businesses and industries are always tempted to ignore or downplay controversies in the hopes that they will just go away. According to Leiss and other business and communications scholars, however, the best strategy for businesses and industries threatened by controversy is to acknowledge the legitimacy of concerns and work quickly and transparently to ameliorate them (e.g., Covello et al. 1989; Fischhoff 1995; Cvetkovitch 1999; Seigrist et al. 2007). This is the so-called Tylenol strategy, a reference to the infamous 1982 product-tampering case, in which manufacturer Johnson & Johnson admitted that its product had been compromised and led the campaign to warn consumers (Snyder and Foster 1983). By taking the communications lead, it is argued, companies and industries can build credibility with stakeholders and the general public, as well as exert a measure of control over the themes, issues, and claims being contested (Powell and Leiss 1997, 36-37). Although the BCSFA was quick to release *Net Gain* as a rebuttal, the document's shadowing of the structure of *Net Loss* and unwillingness to acknowledge the base legitimacy of concerns (even if Kenney and the BCSFA objected to the details) meant that industry supporters could not take the communications lead away from detractors.

The second reason why it is important that *Net Loss* appeared first is that it allowed Ellis to *selectively engage* with the arguments and claims of industry and government. As will be shown using other, more recent cases of conflict over science and fact, this enables critics to choose their points of attack. They are free to set up straw men – to paraphrase opponents and simplify their arguments. In contrast, industry supporters are compelled to *completely engage* with the arguments of their opponents. In ceding the communications lead to critics, supporters of aquaculture are left with much less freedom to choose the terrain over which they are fighting.

The *Net Loss/Net Gain* exchange also illustrates differing choices about language and argument. Both publications are examples of "science communication" in that they present scientific positions and facts using nontechnical language that is understandable to non-expert readers, but there are clear differences between the publications in how facts and claims are presented and discussed. First, *Net Loss* seeks to establish what we will call an "authoritative disinterestedness." That is, its language and structure aim to convince readers that the arguments presented are both authoritative and independent. Several general and specific tools are used in this strategy. In order to establish disinterestedness, opposition groups have often turned to outside authors. The Ellis report was authored by a consultant, and later reports have been written by academics (e.g., Volpe 2001; Krkosek 2005) and even a retired justice of the Supreme Court of British Columbia (Leggatt 2002). Disinterestedness and independence is written into the language of all of these documents. For instance, *Net Loss* begins with the following statement: "Without a broadly based understanding of the industry, it is impossible to consider the impacts of salmon aquaculture ... The purpose of this report is to provide such understanding independent of the various participants and regulators of the industry" (Ellis 1996, 1).

The claim of independence reflects a clear strategic choice. Environmentalist groups are sometimes accused of "Chicken Little-ism," where activists make extreme statements in an effort to galvanize public opinion (Guber 2003, 4). In this sense, the effectiveness of reports such as *Net Loss* comes from the avoidance of extreme language. Instead, "authoritative disinterestedness" is built by pursuing weight-of-evidence arguments. For instance, the Ellis report contains long lists of facts presented in plain language that are well referenced to scientific publications or personal communications with experts. Moral arguments are separated from the presentation of facts, usually appearing in concluding paragraphs, such as the following:

> This industry has grown with little public information or input, and scant consideration for current public costs and the potential for massive losses that would be borne by the biophysical environment and the community as a whole. This outdated model – the use of common property for industrial growth, with narrow private benefit at the expense of society – is unacceptable in British Columbia and Canada today. (Ellis 1996, 145)

Although the moral argument is relegated to the background, its message is clear: by raising these objections, we are acting in the best interests of all citizens. In contrast, the language employed in *Net Gain* is less restrained. While the arguments focus on scientific facts and claims, the accompanying language is often harsh, condemning, and angry. For instance, Kenney (1997, 1-2) argues in the opening paragraphs of *Net Gain* that:

> The Ellis report was flawed from its initial conception in [several] important ways:
>
> • Lack of Academic Rigor: The Ellis report purports to be an academic review but fails to meet the standards necessary to qualify as such by the omission of key scientific reports. The report presents a one-sided view of complex issues based on selective and inadequate research. The intent of the Ellis report is a transparent effort to present the salmon farming industry in an unrealistic and often completely false manner.
> • Bias: The report misrepresents many well-known facts regarding the environmental impacts and regulation of the salmon farming industry in BC. Indeed, much of the extensive literature [on aquaculture] has simply been neglected in the Ellis report to advance the author's narrow views on the issues.

This excerpt raises several important points. First, we see that the factual and moral cases against adversaries are made simultaneously. Again from the perspective of communication theory, this is less desirable, as it conflates evidence and editorial (Covello et al. 1989, 132). It makes the industry appear overly defensive and too ready to lash out at critics (cf. Loftstedt and Renn 1997, 134). Second, and related to this, we see that the moral discourse in *Net Gain* is substantially different. As we saw in the previous quotation, *Net Loss* takes a broadly moral stand (that the aquaculture industry "is unacceptable in British Columbia and Canada today"). This is what Cvetkovitch (1999) calls a "shared values" statement, or an attempt to align the claims maker with the concerns and preoccupations of stakeholders. In contrast, the moral narrative in *Net Gain* revolves around the bias of *Net Loss* and its unfair treatment of the industry. This narrow and self-referential moral argument is unlikely to resonate with concerned members of the general public who will not easily relate to an industry-as-victim narrative.

Numerous conflicts over scientific facts and claims have erupted between aquaculture opponents and supporters since the *Net Loss/Net Gain* exchange. Although the substance of these conflicts has varied, they have unfolded in similar ways. In the rest of this section, we will consider conflicts over escapes, contamination with polychlorinated biphenyls (PCBs), and sea lice.

Escapes

The controversy over escapes is one of the oldest in the industry. Escapes from salmon farms in particular are relatively common, and typically result from equipment failure or bad weather. Escapes are controversial for two reasons: the prospect that escaped fish will interbreed with wild stocks and thus genetically alter them, and the prospect that escaped fish will displace wild fish stocks by competing for habitat and spawning groups. The latter concern is particularly intense in British Columbia, where Atlantic salmon are non-native and thus a source of unknown danger.

Despite the long-simmering nature of the controversy over escapes, anti-aquaculture activists were still able to catch the industry off guard with a series of scientific and activist publications in 2000-01. The lead author for these publications was John Volpe, a respected fisheries ecologist presently based at the University of Victoria, who also has links to some of the most significant anti-aquaculture activist groups in Canada. Volpe has been involved in aquaculture-related research since his days as a doctoral student in the mid-1990s, during which time he participated in a program to monitor British Columbia rivers for evidence of escaped Atlantic salmon. Data were collected primarily by visual observation (snorkelling the rivers). This led Volpe and several colleagues to conclude that at least three river systems in British Columbia contained conclusive evidence of spawning Atlantic salmon (Volpe 2001, 18). In the late 1990s, Volpe constructed an artificial habitat at the University of Victoria to test the ability of aquacultured Atlantic salmon to spawn in a simulated Pacific river system, ultimately concluding that this was indeed possible (Volpe et al. 2001a).

In 2000 and 2001, Volpe and colleagues began publishing findings from these two studies in scholarly journals (Volpe et al. 2000, 2001a, 2001b). The most significant blow to the aquaculture industry, however, came with the publication in October 2001 of the report *Super-Unnatural: Atlantic Salmon in BC Waters,* produced by the David Suzuki Foundation (Volpe 2001). Like David Ellis' *Net Loss, Super-Unnatural* positions itself as "filling the gap" on a little understood issue:

> One of the least understood of the world's major environmental issues is the movement and eventual establishment of species beyond their native range. In contrast to other significant environmental problems such as urbanization and pollution, "biotic invasions" mean that living organisms are the threat ... The transfer of organisms [across] environments is rapidly eroding worldwide biodiversity and reshaping Earth's natural systems in our own lifetime. While the rest of the world scrambles to address the challenges presented by exotic species, Canada – a country with more to lose than most – has remained silent. (Volpe 2001)

Super-Unnatural goes on to compare the use of Atlantic salmon in Pacific waters with disastrous cases of introduced species, such as the Australian rabbit crisis and the plague of zebra mussels in the Great Lakes (Volpe 2001, 5-6). The main body of the report, however, is dedicated to challenging "the politically palatable responses of government and industry" to the problem of escapes. This discussion profits from the key advantage that was also seized by Ellis in *Net Loss:* the ability to selectively engage with the claims of adversaries. For instance, the "politically palatable responses of government and industry" that Volpe (2001, 12) argues against are:

- That escapes [of Atlantic salmon] are very rare
- That escapes of Atlantic salmon are inevitable but they won't survive in the wild
- That some Atlantic salmon may survive but will not ascend freshwater rivers
- That some adult Atlantic salmon are likely to be found in freshwater rivers but can't spawn
- That spawning is likely to occur but progeny will not be competitively viable
- That multiple-year-classes of juvenile Atlantic salmon in some rivers do not pose a threat to native populations

Volpe's dissection of these claims is thorough and factual. It is important to recognize, however, that the claims quoted above are themselves never referenced or attributed. This does not mean that "the politically palatable responses" outlined by Volpe are fabrications. Having followed the aquaculture controversy for some time, we have encountered each of these arguments against the probability that Atlantic salmon will spawn and colonize Pacific waterways. Our point is that, as with Ellis' *Net Loss,* the author is selectively engaging pro-aquaculture claims, voices, and narratives. By speaking first on the issue (in a public sense), authors are free to summarize, paraphrase, and choose points of attack.

As was the case following the publication of *Net Loss,* industry supporters quickly responded to *Super-Unnatural* with publications of their own, most notably a document entitled *On the Risk of Colonization by Atlantic Salmon in BC Waters,* released in May 2002 by the BCSFA (Ginetz 2002). The report, authored by DFO biologist Ron Ginetz, adopts the same "rebuttal and discreditation" strategy as Kenney's *Net Gain.* It rebuts Volpe's warnings about colonization by highlighting the historical failure to introduce Atlantic salmon to the Pacific coast despite repeated and concerted efforts by the federal government and anglers' groups hoping to establish a sport fishery. Ginetz points to the fact that "from 1905 to 1935 in excess of 8.6 million Atlantic salmon were intentionally introduced to more than 60 British Columbian lakes and streams," but that no population was established (Ginetz 2002, 3). As with *Net Gain,* the body of the Ginetz report is dedicated to countering specific claims made in *Super-Unnatural* and in Volpe's scholarly publications. In this case, the rebuttals are organized in claim/fact dichotomies, such as:

Claim: Colonization is already occurring, or has already taken place as evidenced by the presence of juvenile [Atlantic salmon] in B.C. rivers.
Fact: The presence of juvenile Atlantic salmon in three B.C. streams does not represent colonization. (Ginetz 2002, 24)

As with the *Net Loss/Net Gain* exchange, therefore, while opponents of aquaculture are free to selectively engage the arguments of their adversaries, the rebuttal response compels industry supporters to completely engage with anti-aquaculture claims. The late entry of the industry into the public discussion about escapes and colonization means that it is forced to follow the debate rather than lead it. Moreover, the Ginetz report, like *Net Gain,* fails to separate the factual and moral cases. Although Ginetz presents a strong statistical and scientific case against colonization, assertions of fact are intertwined with strong attacks on environmentalists: "Colonization by Atlantic salmon in coastal waters of the Pacific Northwest has been examined by several authorities over the past decade [4 sources cited], and all have concluded that the risk of colonization is low. Nevertheless, and whether from ignorance, mischief or other self-serving agendas, some continue to argue against balanced ecological assessments with unscientific arguments"

(2002, 9). As we have seen, this mix of assertion and discrediting, evidence and editorial, damages the credibility of the overall argument in the eyes of third parties (cf. Covello et al. 1989; Cvetkovitch 1999).

PCBs and Sea Lice

The same broad patterns were repeated in controversies erupting over PCB contamination of aquaculture products and the issue of sea lice transfer from aquaculture operations to vulnerable migrating juvenile wild salmon. In their study of the PCB conflict that occurred in early 2004, Leiss and Nicol (2006, 901) conclude that "the controversy should not have surprised the government or the fish farming industry." They point out that three previous studies published in academic journals reported elevated levels of the compounds in aquacultured fish compared with wild stocks (Easton and Luszniak 2002; Jacobs et al. 2002a, 2002b). Again, however, in January 2004 the industry was rocked by a major communications offensive by anti-aquaculture groups. This offensive began with the publication in the top-flight academic journal *Science* of an article by Indiana University chemist Ron Hites and several colleagues, entitled "Global Assessment of Organic Contaminants in Farmed Salmon" (Hites et al. 2004). The article itself would probably not have garnered much media attention, but its publication was accompanied by press conferences, media releases, and a website launch bankrolled by major environmentalist groups in the United States (Hoijer et al. 2006, 278). The strategy worked brilliantly. In a study of global media coverage of the PCB controversy, Hoijer and colleagues (2006) catalogued seventy-four articles that appeared in major newspapers in Europe, North America, and South America in the two weeks after the Hites article appeared. Our own research shows that coverage in Canada was particularly strong, with thirty-seven articles in major daily newspapers. The salmon aquaculture industry was severely affected by this controversy, with some sources indicating that global sales of farmed salmon dropped by as much as 70 percent just days after the first media coverage (Leiss and Nicol 2006, 900).

Industry supporters again reacted by releasing rebut-and-discredit materials. The most thorough of these was a report commissioned by Positive Aquaculture Awareness, *Framed Salmon: Farmed Salmon, PCBs, Activists, and the Media* (Moore 2004).[7] The structure of *Framed Salmon* is similar to that of the Kenney and Ginetz reports, in that the main body of the document is dedicated to a technical critique and rebuttal of the scholarly work linking aquaculture to high PCB contamination. Again, however, arguments about evidence are intertwined with attacks on opponents. Indeed, *Framed Salmon* opens with the statement: "As you will see from the following report ... we found surprising evidence of contrived information and irresponsible journalism. It seems clear that these findings [of PCB contamination] form part of the larger effort by activists to damage the reputation of the salmon aquaculture industry by using food-scare tactics that have no basis in scientific fact" (Moore 2004, ii).

The storyline of the sea lice controversy is also strikingly familiar. Concerns had been voiced since the 1980s about the incubation of diseases and parasites on aquaculture farms, as well as their potential transfer to wild populations (Keller and Leslie 1996, 31). Sea lice are naturally occurring marine parasites (*Lepeophtheirus salmonis* and *Caligus clemensi*). At early life stages, lice are free-swimming, meaning that they float in ocean currents, awaiting contact with hosts. Once attached, sea lice feed on fish flesh and mucus, and typically fall off in winter months. Given this life cycle, normal sea lice infections are relatively harmless to infected hosts. They are, however, a significant problem for aquaculturists due to the high population densities of farm sites, and severe outbreaks are sometimes treated using pesticides (Westcott et al. 2004).

In 2001, the sea lice issue took a dramatic turn as biologist and environmental activist Alexandra Morton reported finding extreme numbers of infections among juvenile wild salmon migrating from spawning rivers (Morton 2004). Controversially, she and others later argued that these infections were responsible for a "population crash" in pink salmon returns to rivers and streams in British Columbia's Broughton Archipelago in 2002, when the number of pinks returning to spawn fell to their lowest levels since 1960 (in even-numbered years, which is significant because of pink salmon's two-year life cycle). Morton and others asserted that sea lice infection had killed the vast majority of juvenile pink salmon once they left spawning rivers and passed aquaculture farms on their way to the open Pacific (Morton 2004, 227).

These findings caused a stir in 2002 and 2003. The DFO conducted its own research (Beamish et al. 2004, 2005) and concluded that sea lice posed little risk to wild salmon populations. The David Suzuki Foundation also funded a study that sampled juvenile wild fish in waters free of salmon aquaculture south of Prince Rupert in order to establish a comparative baseline for Morton's Broughton studies (Rolston and Proctor 2003). In February 2003, the government of British Columbia ordered the temporary closure of several salmon farms in the Broughton Archipelago as part of a precautionary approach to protect the upcoming year's juvenile runs (British Columbia 2003b).

Despite the fact that the controversy had brewed for years, the salmon aquaculture industry in British Columbia was again put on its heels by a concerted media campaign timed to coincide with new scholarly publications on sea lice transfer. In 2005, a group of scholars (including John Volpe and Alexandra Morton) co-authored and published a series of articles in prestigious journals (including the *Proceedings of the Royal Society of London*), arguing that mathematical modelling of juvenile salmon movements in the

Broughton Archipelago proved that aquaculture sites were the source of sea lice infection (Krkosek et al. 2005a, 2005b). As with the PCB controversy, these publications were accompanied by companion materials aimed at media and the general public. The David Suzuki Foundation and the Raincoast Conservation Society released a series of pamphlets and press releases that summarized the publications and outlined "implications" (DSF-RCS 2005). As with *Net Loss*, the two organizations' joint media release pursued a weight-of-evidence approach followed by conclusions stressing broad moral issues:

> These data, due to the massive sampling effort and the unequivocal nature of the conclusions, satisfy even the most conservative benchmark for proof – this is the definitive work on the issue ...
>
> Given the increasing weight of evidence, including this research, that salmon farms are a potentially lethal source of sea lice on wild salmon, there is ample evidence to compel precautionary action by regulators. The premise of industrial-scale open net aquaculture in wild salmon habitats needs to be reconsidered ... There is a clear potential for severe and irreversible damages to be inflicted upon wild salmon populations and their dependent cultures, ecosystems, and economies. (DSF-RCS 2005, 3-4)

This strategy succeeded in attracting mainstream media coverage, and twelve news items appeared in major Canadian newspapers in the week after the press release. Soon after, industry supporters released a series of critiques and rebuttals (Aquaculture Association of Canada 2006; BCSFA 2006; Brooks 2006; Butterworth et al. 2006; CAIA 2006b). Brooks (2006) offered a technical critique of the Krkosek papers (Krkosek et al. 2005a; 2005b), while the BCSFA (2006, 2) argued that "the Krkosek paper seems to be very 'pompous' without having the results to verify their claims," and that "the introduction [of the paper] is quite inflammatory and not balanced." In this case, as in earlier controversies, the industry and its supporters assumed a reactive stance: advancing a narrow moral argument, focusing strongly on rebutting their adversaries' words, and thus ultimately failing to defuse the controversies.

Conclusion

In this chapter, we have argued that the aquaculture controversy is intertwined with the changing politics of knowledge. For most people, aquaculture is learned rather than lived. This is not unusual in a knowledge society, and the same can be said of controversies such as nuclear power generation, the genetic modification of organisms, and agricultural pesticide use. Most people encounter these only at the end point, on our plates, through our utilities, and so on. We know about them primarily through knowledge claims rather than first-hand experiences.

Like nuclear power (Gamson and Modigliani 1989) and genetic modification (Leiss 2001), aquaculture was relatively uncontroversial in its early days. It has become controversial through the actions of claims makers who are increasingly able and adept at mobilizing both science and narrative, certainty and uncertainty, to establish alternatives to official claims. The rise to prominence of newer "holistic" disciplines such as ecology, as well as the diffusion of scientific language and expertise away from traditional homes and authorities (university, government, industry), have enabled activist groups to directly and indirectly produce scientific knowledge and fact. Moreover, the pluralization of science has also made the general public more receptive to alternative scientific and expert claims. There is now a cultural expectation, in Canada and in other advanced democratic societies, that science and scientific processes be more transparent and inclusive – that they "come down to earth" and recognize public concerns and values.

Critics of aquaculture have taken advantage of these changing expectations. They have also been strategically successful in advancing interpretive frames (such as taint and risk) that resonate in the public sphere. In contrast, the industry and its supporters have thus far struggled to assert their points of view in this new arena. Opponents of aquaculture have steered public debate by assuming an offensive stance (which gives them the capacity to summarize, paraphrase, and choose points of attack) and using broad moral arguments based on weight-of-evidence logic. In each of the cases considered above, industry supporters have responded to communications offensives with rebut-and-discredit materials that too often blend evidence and editorial (including personal attacks on opponents) and that fail to acknowledge the validity of critiques and skepticism of aquaculture (even if they disagree with the claims and interpretations presented). According to the growing literature on risk issue management, this approach does not resonate well with concerned members of the general public, who expect these concerns to be seriously and transparently addressed (Covello et al. 1989; Leiss 2001; Seigrist et al. 2007; Chryssochoidis et al. 2009). As argued by Slovic (1999), the knowledge battlefield is as much about trust and emotion as it is about fact. Scientific claims are central to the aquaculture conflict, but "more and more science" alone will not necessarily move us closer to consensus or resolution on this issue.

4
Knowledge Warriors? Experts and the Aquaculture Controversy

If the aquaculture controversy in Canada is a knowledge battlefield – where pro- and anti-aquaculture activists publicly struggle over whose claims and interpretations are most valid – then it is important to look carefully at the claims makers. In Chapter 3, we looked at the end of the pipe of the knowledge conflict, where interested groups strategically deploy facts and narratives in public efforts to legitimize their positions and advance political goals. In this chapter, we examine how scientists and other experts in aquaculture understand the controversy and their own roles within it. We do this by presenting findings from a survey completed by 300 aquaculture experts across Canada – from universities, government, industry, and environmentalist organizations. This survey gives us a window into the professional lives and personal opinions of experts on all sides of the aquaculture debate.

Findings from the survey indicate that experts on aquaculture are deeply divided. There is little evidence of consensus on the risks and benefits of aquaculture. Instead, there is a broad spectrum of professional opinions, from experts who are strongly convinced of the desirability of aquaculture, to those who are firmly against it, to moderates of every degree who are somewhere in the middle.

This chapter has several aims. The first is to search for patterns by looking closely at the disagreements among experts. For instance, data from the survey show that opinions on aquaculture are strongly associated with differences in experts' institutional affiliations, career lines, professional networks, and values and beliefs. These findings are controversial, because they can imply that experts' professional opinions on aquaculture may be explained in large part by personal and social variables (cf. Longino 1990). We avoid drawing firm conclusions on this point, but we *do* argue that the data suggest that experts' professional and personal situations can contribute to particular "lenses" through which they see the issues.

Second, we look at the consequences of the controversy for trust within the expert community. Science depends on trust – scientists and experts need to trust the work of others in order to build upon it (Hardwig 1991). Using closed- and open-ended data from the survey, we find, unfortunately, that mistrust is a significant problem for experts on all sides. Many respondents expressed deep skepticism of the motives and methods of those who disagreed with them, and this does not bode well for future consensus or moderation on aquaculture. At the same time, however, we find that some experts reject the notion of mistrust outright, arguing instead that problems of bias emerge in the application of scientific knowledge by outsiders. As we shall see, this defence-of-science argument is a recurring theme in experts' understandings of the controversy.

Third, we examine experts' views of public participation in the controversy. As discussed in Chapter 3, the public is more and more involved in scientific debates, particularly over questions of environment, health, and industrial development. Research dealing with other controversies tells us that experts are conflicted about this (e.g., Wynne 2001). Our survey finds that aquaculture experts on all sides of the debate are open to public participation but are highly critical of the public's ability to arrive at responsible conclusions about aquaculture.

Overall, our survey paints a complex portrait of aquaculture experts in Canada, who are alternately stubborn and reflective, hostile and conciliatory. At times, we can see real divisions in the expert community, where battle lines are clearly drawn and experts line up on either side of an issue. At other times, we find that experts on all sides share similar perspectives, particularly on questions of trust and views of the public. At those moments of convergence, aquaculture experts are far from being "knowledge warriors," but rather are frustrated people trying to make sense of complicated situations.

The Survey

As we saw in Chapter 3, moderate voices are easily lost in arguments about aquaculture. The public voices and faces of aquaculture science tend to represent the poles of the debate. These extreme voices are important because of their significant impact on public and political discourses about aquaculture (we will examine media coverage of aquaculture issues in Chapter 5). It became clear early in our research, however, that many aquaculture scientists and other experts hold moderate and nuanced views of the industry and its effects. Far from being "knowledge warriors" engaged in crusades for and against the industry, in informal conversation most of them came across as reflective, thoughtful, and uneasy with the tenor of the public and scientific debates over aquaculture in Canada.[1]

One of the main goals of the survey, therefore, was to capture the diversity of expert opinion on aquaculture that is often lost in activist narratives for and against the industry. In determining our sample population, we attempted to locate *all persons in Canada* with a claim to some form of science-based expertise or authoritative knowledge with respect to aquaculture. We used two general criteria in selecting potential participants: (1) the person has direct involvement in aquaculture research and/or the communication of research findings, *and* (2) the person has a formal educational background in natural and/or social sciences.[2]

There is now a significant body of literature that advocates a rethinking of "expert" and "lay" as distinct categories (cf. Levy-Leblond 1992; Collins and Evans 2002). For instance, concerned citizens often become "experts" on specific issues despite a lack of formal scientific training (Tesh 1999). Alternatively, some authors argue that lived experience or local knowledges are forms of expertise that ought to be placed on par with science-based expertise (cf. Yearley 2000). These arguments are valid. Nevertheless, we saw in Chapter 3 that the knowledge battlefield for aquaculture is contested mainly on questions of science-based facts. Given the influence of science in this issue, we consider the traditional definition of expertise to be the most useful way to define our population.

Our survey population was built using information in the public domain. The most common method was to cross-reference employment directories with job titles, individual biographies (such as personal Web pages), and records of publication. We sought to include participants from a wide variety of institutional backgrounds, including academia, government, industry, environmentalist groups, and Aboriginal organizations. Using these methods, we invited 502 individuals to complete the survey. Three hundred responded, a response rate of 60 percent (most self-administered questionnaires achieve a response rate of between 20 and 35 percent; see Kaplowitz et al. 2004, 98).[3]

These methods enabled us to hear from the "knowledge elite" on aquaculture issues in Canada. Among respondents, 60 percent held PhD, MD, and/or DVM (Doctor of Veterinary Medicine) degrees; 20 percent indicated MSc as highest degree attained; 17 percent held a BSc degree; and 4 percent were otherwise certified.[4] Moreover, 82 percent of respondents indicated some direct involvement in original research. Thus, while we use the term "experts" instead of "scientists," respondents were generally well educated and directly involved in the production and/or handling of science-based claims.

Table 4.1 gives a rough indication of the response rates across the different institutional affiliations of aquaculture experts. Column two presents the total population that we expected, based on our research in building the sample population, while column three presents self-reported data – what respondents themselves indicated as their institutional affiliations. Based on this rough comparison, response rates were highest among experts affiliated

Table 4.1

Institutional affiliation of survey respondents compared with expected affiliations based on population sample

Institutional affiliation	Expected (from background research)	Actual (as reported by respondents)	Response rate (rough %)
University or college	181	103	57
Government	259	138	53
Industry	31	27	87
Environmentalist group	18	15	83
Aboriginal group	8	4	50
Other	11	13	118
Total	508	300	60

with industry and environmentalist groups, and lowest among experts affiliated with governments and Aboriginal organizations. One of the regrettable shortcomings of the survey is the low number of respondents affiliated with Aboriginal organizations. As mentioned, we built the survey population largely by means of information in the public domain, including Internet sites and employee directories, and it is possible that many Aboriginal organizations do not engage in such public listing of employment information. In any case, the low number of respondents affiliated with Aboriginal groups meant that, for the purposes of statistical analysis, we incorporated this category into "Other."

The survey was conducted between May and September 2002. This overlapped with the first round of major scientific and public debates regarding the possible transfer of sea lice from aquaculture operations to migrating juvenile salmon in British Columbia's Broughton Archipelago (Morton 2004), as described in Chapter 3. Thus, the survey provides a snapshot of the scientific and expert community at a very intense moment in the aquaculture controversy. It is unclear whether this snapshot effect had any impact on our data, for instance by heightening respondents' emotions or frustrations. It is clear from the discussion in Chapter 3, however, that these moments of crisis and turbulence are common – meaning that the snapshot was not necessarily taken during an exceptional time but rather a time of particularly intense conflict involving scientific claims.

Experts' Assessments of Aquaculture
One of the primary goals of the survey was to measure the diversity of opinions and assessments of aquaculture within the expert community. Capturing an expert's assessment or stance regarding aquaculture presented a challenge,

however. As discussed in earlier chapters, the aquaculture controversy involves multiple fronts or axes of disagreement, specifically over environmental integrity, human health, rights, and rural development. This multidimensionality is important, as it is conceivable (and quite common) for experts to hold different views across these dimensions. To capture this, we constructed a fourteen-item scale as our measure of experts' assessment or stance regarding aquaculture (referred to henceforth as the "Stance Scale"). Each item in the scale involves a pairing of two incommensurable statements that address a particular axis of the controversy. The scale items are given in Figure 4.1. The statements contrast assessments of harm or risk associated with aquaculture with assessments of benefits or the absence of risk. Respondents were asked to indicate which of the paired statements best represented their opinion (along with the options of "neutral or undecided" or "don't know").

The paired-statement method is an established means of exploring competing views on complex issues (Kallerud and Ramberg 2002, 220). As evident from Figure 4.1, some statements are specifically "expert" in nature, asking participants to draw on technical or scientific knowledge in considering their responses. We conducted pre-tests with graduate students in aquaculture programs at two Canadian universities to ensure that the statements were not *too* technical and would be familiar to potential participants. During debriefing, these pre-test participants indicated a strong familiarity with each of the items in the scale. Meanwhile, other statements are less technical, particularly those that address issues of rural economic development and Aboriginal rights. We argue that statements such as "Coastal communities are harmed by aquaculture" and "Coastal communities benefit from aquaculture" are important, because this is potentially a key dimension of experts' stance on aquaculture. Also, it is important to note that experts, like the general public, do not refrain from developing perspectives on issues outside their immediate area of expertise.

Overall, the survey yielded several striking patterns with respect to these statements. First, there is substantial disagreement on nearly every item. For statistical purposes, we coded responses as follows: agreement with the risk/harm statement was given a score of –1, "neutral or undecided" a score of 0, and agreement with the benefit/absence of risk statement a score of +1.[5] When all responses for each paired statement are averaged, a score approaching 1 indicates consensus regarding a positive aspect of aquaculture, a score approaching –1 indicates consensus regarding a negative aspect of aquaculture, and an average score near 0 indicates disagreement.[6] Only three statements yielded anything approaching consensus: (1) "Aquaculture is a major economic assistance to local businesses" (mean of 0.73); (2) "Aboriginal people have a lot to gain by being involved in aquaculture" (mean of 0.69); and (3) "Coastal communities benefit from aquaculture" (mean of 0.66).[7]

Figure 4.1

Stance Scale items used to measure experts' assessment of aquaculture

Assessment of harm or risk	Assessment of benefit or absence of risk
Coastal communities are harmed by aquaculture	Coastal communities benefit from aquaculture
Finfish farming is a major source of pollution	Finfish farming is NOT a major source of pollution
Shellfish farming competes with traditional ocean uses	Shellfish farming is consistent with traditional ocean uses
Aquaculture has a negative effect on local businesses	Aquaculture is a major economic assistance to local businesses
Aquaculture does damage to the traditional fisheries of Aboriginal people	Aquaculture has NO impact on the traditional fisheries of Aboriginal people
Aboriginal people have a lot to lose from being involved in aquaculture	Aboriginal people have a lot to gain by being involved in aquaculture
Aquaculture constitutes a threat to wild fish stocks	Aquaculture serves to reduce the pressure on wild fish stocks
Aquaculture poses a threat to human health	Aquaculture poses NO threat to human health
The presence of aquaculture exposes wild fish to a heightened risk of disease	The presence of aquaculture does NOT expose wild fish to a heightened risk of disease
Farmed fish are more likely to spread diseases to wild fish than the other way around	Wild fish are more likely to spread diseases to farmed fish than the other way around
Aquaculture is likely to bring about a loss of genetic diversity in fish stocks	Aquaculture is unlikely to have any effect on the genetic diversity of fish stocks
The use of therapeutic drugs in aquaculture constitutes a threat to human health	The use of therapeutic drugs in aquaculture does NOT constitute a threat to human health
Research involving genetic modification in fish constitutes a potential threat	Research involving genetic modification in fish constitutes NO threat
The escapes of farmed fish will likely lead to the displacement of wild fish stocks	Escapes of farmed fish will NOT likely lead to the displacement of wild fish stocks

Importantly, each of these refers to positive evaluations of local economic issues. In contrast, the most controversial statements (with scores near 0) refer to environmental issues. For instance, the paired statements "The presence of aquaculture exposes wild fish to a heightened risk of disease" and "The presence of aquaculture does NOT expose wild fish to heightened risk of disease" achieved a mean score of –0.11, and the statements "Aquaculture is likely to bring about a loss of genetic diversity in fish stocks" and "Aquaculture is unlikely to have any effect on the genetic diversity of fish stocks" achieved a mean score of 0.04.

The true value of these measures is revealed when we combine all fourteen items into the Stance Scale. Given that each item is scored from −1 to +1, summing the items together gives us an overall scale ranging from −14 to +14. There are two main statistical methods for testing the validity of a scale. The first is a measure of reliability, or the ability of a scale to produce a consistent measure across all respondents, called a *Cronbach's alpha test*. On a reliable scale, an individual's answer to one question should predict (though not perfectly) how he or she will respond to the other questions that make up the scale (Cronbach 1951). A scale is considered to be valid if the Cronbach's alpha value exceeds .70 out of a possible 1.0. Our scale achieves a score of .90.

The second method for testing the validity of our Stance Scale involves factor analysis. Whereas the Cronbach's alpha tests for the tendency of scale items to vary together, factor analysis looks for overlap among the items. If two or more scale items overlap too strongly (if, for instance, many respondents are answering the questions in Figure 4.1 in the same way), then including them in the scale risks double-counting what are essentially the same opinions or dimensions. Factor analysis helps to avoid this by measuring the contribution of each item to the overall variance in the scale. Factor analysis assigns each item a score from 0 to 1.0. For a population of our size (300 respondents), a score of .40 or higher indicates that the item makes an original contribution to the scale (Blaikie 2003, 221-22). All fourteen items scored at or above .40, with eleven items coming in above .60. In summary, the statistical tests show that our Stance Scale for aquaculture achieves the best of both worlds. The Cronbach's alpha test shows that the items in Figure 4.1 tend to vary relative to each other (meaning that the scale consistently measures a singular phenomenon), while factor analysis reveals that each item measures a different dimension of experts' overall assessment or stance on aquaculture.

Patterns in Expert Assessments of Aquaculture

The next step is to look for patterns in experts' assessments or stance regarding aquaculture. We hypothesize that experts' stance on aquaculture is likely to vary according to three criteria. First, we expect that there will be substantial differences across the institutional affiliations of universities, governments, industry, and environmentalist groups. We expect the strongest differences between the last two groups, with weaker effects in the more diverse institutions of academia and government. Second, we expect that experts' stances will vary across what we shall call "individual situations." This is a complicated notion. On the one hand, we expect that *personal* variables, such as gender, education, and nativity (birth in or outside of Canada), might relate to opinions on aquaculture. The relationship of gender to science, for instance, remains controversial. Some feminist authors argue that scientific knowledge has been built on ideologies of masculinity. These ideologies, which give priority to ideals such as rationality, objectivity, careerism, and the segregation of science from other activities and types of knowledge, may introduce a gender bias into the methodologies and cultures of science (cf. Rose 1983; Haraway 1988; Harding 1991). It is argued that women, holding different values and standpoints, are less bound by such ideologies and thus have different perspectives on scientific knowledge and authority (see Smith 1987). An expert's individual situation also involves professional networks and activities. Network theories in sociology tell us that interpersonal connections are key channels for the flow of knowledge, values, and influence (Collins 1974; Podolny et al. 1996), and we therefore expect that experts' professional networks and experiences may relate to their assessment of aquaculture. Third, we expect to find a relationship between values and stance on aquaculture. The link between science and values is controversial, particularly because value-neutrality is an important part of the ideology of science and its claim to authority (see Chapter 3). Nevertheless, scientists are human beings, and humans use values to interpret knowledge (Longino 1990). We therefore expect that different stances on aquaculture may be related to differences in personal values.

Before we begin looking at the data, it is important to clarify what we are seeking to find. First and foremost, we are looking for patterns. An expert's stance on aquaculture is a complicated thing, and we are looking for trends within this complexity. In doing so, we are not looking to *explain* why some experts support aquaculture while others oppose it. Aquaculture experts have spent a long time developing an understanding of this issue, and this process cannot be captured using survey methods. What we will argue, however, is that aquaculture experts approach the issue using different lenses, and that these lenses are strongly linked to professional and individual situations – that is, these different lenses are in large part built from different professional and personal experiences, and aquaculture looks different depending on these experiences. This is an ambitious argument but, as we shall see, the survey has uncovered some strong patterns in how experts view the aquaculture industry and its effects.

Institutional Affiliation

Table 4.2 presents the average scores on the Stance Scale achieved by experts affiliated with different institutions. As discussed earlier, the Stance Scale runs from −14 (a very negative view of aquaculture across each dimension outlined in Figure 4.1) to +14 (a very positive view). Table 4.2 shows that

Table 4.2

Survey respondents' stance on aquaculture according to institutional affiliation

Institutional affiliation	Mean Stance Scale scores[a]	Standard deviation
University or college	+2.3	6.2
Government	+5.0	6.0
Industry	+9.6	4.4
Environmentalist group	−3.8	9.6
Other	+2.5	7.5
Total	+3.9	6.7

a   The Stance Scale runs from −14 to +14.
Measure of association: *eta*-squared = .16
Analysis of variance: $F = 12.4$; $p < .001$

experts affiliated with industry have the strongest and most consistent views on aquaculture, with an average score of +9.6 and a standard deviation (the average deviation from the mean) of 4.4 points on the scale. The category of experts who are most critical of aquaculture are those affiliated with environmentalist groups, which had an average score of −3.8. This score is closer to the midpoint of the scale than we expected, but the high standard deviation of 9.6 points on the scale indicates significant variance within this category of experts. Closer examination of this group reveals that much of this variance is driven by several outliers, or a small number of respondents who indicated affiliation with environmentalist groups and held positive views of aquaculture. When these outliers are excluded, the mean score for experts affiliated with environmentalist groups drops to −7.3 and the standard deviation to 6.4.[8] This standard deviation is still quite high, which suggests that there is greater diversity of opinion in environmentalist groups than is suggested by some of the activist materials and publications discussed in Chapter 3.

Respondents affiliated with universities or colleges had moderate average scores on the Stance Scale (+2.3). This is probably because universities and colleges are diverse institutions that house a wide variety of experts and expert opinions. The high standard deviation of 6.2 points on the scale reflects this diversity. Government departments and agencies also employ a wide variety of experts, but their mean scores show a generally positive stance (+5.0), albeit with a high standard deviation of 6.0. Finally, the category "Other," which includes independent consultants and researchers as well as experts affiliated with Aboriginal organizations, scored a moderate +2.5, with a high standard deviation of 7.5.

The high standard deviations in Table 4.2 tell us that significant differences of opinion exist within the different categories. Nevertheless, statistical tests show that institutional affiliation is still a major predictor of a respondent's assessment or stance regarding aquaculture. Because we are comparing a nominal variable (affiliation) to an interval variable (the twenty-eight-point Stance Scale), we can use the *eta*-squared statistical test to tell us how much of the total variance on the Stance Scale can be explained by knowing respondents' affiliations (Blaikie 2003, 134). The test yields an *eta*-squared value of .16, which means that 16 percent of the total variance – which represents people's different opinions of aquaculture – can be predicted by knowing where respondents work.[9] To be clear, we do *not* argue that one is causing the other, only that there is a relationship. In all likelihood, the relationship is multi-causal: people choose to work in an environment that is consistent with their views, and this environment then reinforces those views. To this point, we can say with confidence that where an expert works is strongly associated with his or her stance on aquaculture, although much of this effect appears to be driven by the "interested" organizations of industry, government, and environmentalist groups.

Individual Situation

Next, we look at variables that extend beyond affiliation to what we call experts' individual situations. An expert's individual situation is made up of personal and professional variables. On the personal side are variables such as gender, education, and nativity (measured as birth in or outside of Canada). With respect to gender, 72 percent of respondents were male and 28 percent were female. In terms of education, we make a distinction between respondents who have earned a research-based doctorate (PhD) and those who have not. In this we anticipate that advanced education in a research-intensive environment may give nuance to experts' views of aquaculture relative to those educated differently. For the purposes of this study, we consider Doctors of Medicine and Doctors of Veterinary Medicine as *not* having earned a research-based doctorate. This leaves us with 47 percent of respondents with research-based doctorates and 53 percent without. With respect to nativity, a sizable proportion (27 percent) of respondents indicated birth outside of Canada. The potential effects of nativity on assessments of aquaculture are not immediately obvious. As discussed in Part 1, aquaculture touches on several major Canadian preoccupations, such as resources and environment, urban/rural relationships, and Aboriginal rights. It is therefore possible that the nativity variable may expose differences between experts born and socialized in Canada and those who arrived later in life.

Professional variables are also potentially significant. The survey contained several measures of professional activities and networks, including employment history and collaborations with other experts. Employment history was measured by simply asking about the respondent's previous employer (if any) according to the same institutional categories discussed above (academia, government, industry, environmentalist group, Aboriginal organization, or other). Professional networks were measured using a "collaboration index." This index was compiled by asking respondents to indicate whether they had done any of the following with colleagues across institutional categories in the past five years: shared data, co-authored papers, co-authored research proposals, and/or collaborated in any other way. Positive indications of each activity were then summed to give respondents a score (0 to 4) for each institutional category. Together, employment history and the collaboration index provide a rough measure of an individual expert's range of connection and contact with other experts across different institutional settings.

In what follows, we apply these measures of individual situation to what we already know – namely, that experts' stances on aquaculture vary by institutional affiliation but that there remain significant differences of opinion within these categories, as shown by the high standard deviations in Table 4.2. Our first step is to see whether the individual situation variables can tell us more about the differences of opinion within the different institutional categories of academia, government, industry, and environmentalist groups. To do so, we use multivariate regression analysis, which is similar to the *eta*-squared test but which enables us to input multiple independent variables *and* to assess their relative importance in predicting experts' opinions on aquaculture (Knoke and Bohrnstedt 1991, 279). The findings of the regression analysis are presented in Table 4.3.

The first row below the headings in Table 4.3 indicates that the personal and professional variables can explain 36.5 percent of the variance in the Stance Scale among university- or college-based experts. Recall that this group scored a moderate average of +2.3 on the scale. The fact that 36.5 percent of the variance within this category can be explained tells us that there are strong patterns among experts whose views of aquaculture scored above and below the group mean of +2.3. The last column of Table 4.3 indicates which variables are providing this explanation, in the order of their importance. For example, "previous employment in industry" is the most significant predictor of deviation from the mean among academic experts, followed by "collaboration with colleagues affiliated with industry" and "collaboration with colleagues affiliated with environmental groups" (we will discuss the direction of these trends in a moment).

The second row below the headings in Table 4.3 shows that personal and professional variables are less capable of predicting different opinions on aquaculture among government experts. The next row, however, shows that nearly 36 percent of the variance among industry-affiliated experts can now be explained. Unfortunately, our sample of environmentalist

Table 4.3

Ability of individual situation variables to explain variance within survey respondents' institutional categories (by stance on aquaculture)

Institutional category	Percentage of variance explained within institutional category[a]	Significant variables (in order of importance)[b]
University or college	36.5	Previous employment in industry; collaboration with colleagues affiliated with industry; collaboration with colleagues affiliated with environmental groups
Government	8.6	Previous employment in a university; collaboration with colleagues in universities; collaboration with colleagues in environmentalist groups
Industry	35.8	Previous employment with government; gender
Environmentalist group	Category too small for meaningful analysis	–

a   The percentage of variance explained is the adjusted R^2 value produced using multivariate regression. The dependent variable in this regression is the Stance Scale, against the independent variables listed in the rightmost column.

b   Each variable demonstrates significance at $p < .05$ (t-test). Order of importance or significance is determined by the beta scores achieved in the multivariate regression. Beta indicates the degree of influence a variable has on the final regression model (Knoke and Bohrnstedt 1991, 279).

nongovernmental organization (ENGO) respondents is too small for us to meaningfully use individual situation to analyze differences in stance on aquaculture.[10]

As mentioned, the last column of Table 4.3 is important because it indicates *which* variables are associated with differences of opinion. Note that *personal* variables are largely absent. No significant differences were attributable to education or nativity. Gender is also an insignificant factor among experts affiliated with universities or government. Only among industry-affiliated experts do women hold significantly more moderate opinions on aquaculture than their male counterparts, when other variables are held constant.[11]

In contrast, the *professional* variables are very significant for predicting differences in opinion within the different institutions. This is particularly

Table 4.4

Percentage of variance on the Stance Scale explained by institutional affiliation and individual situation, based on a two-step regression model

Step	Variables considered	Percentage of total variance explained
1	Institutional affiliation	16
2	Institutional affiliation + individual situation	28

true of universities and colleges. Academics who have professional links to industry-affiliated experts are significantly more supportive of aquaculture than those who do not, whereas those with ties to environmentalist groups are generally more critical.[12]

Next, we look at the total effects that institutional affiliation *and* individual situation have on different stances on aquaculture. We do this by performing a two-step multivariate regression analysis. As shown in Table 4.4, the first step involves re-creating the finding that we discussed earlier: that institutional affiliation alone can account for 16 percent of the total variance on the Stance Scale. The second step involves introducing the individual situation variables into the regression (alongside institutional affiliation). Table 4.4 shows that introducing information about a respondent's personal and professional background nearly doubles the power of the model to explain variance in expert opinions. When considered together, the two sets of variables can account for 28 percent of total variance.[13]

These findings demonstrate that conflicting expert opinions on aquaculture are distributed in large part along institutional and network lines that are statistically predictable. To reiterate, this does not mean that these variables *cause* experts to reach certain conclusions about aquaculture. It does indicate, however, that expert opinions on aquaculture are associated with different career activities and networks. Personal variables such as education, gender, and nativity did not prove to be significantly linked to opinions on aquaculture (with the notable exception of gender among industry experts). On the other hand, the measures of professional variables are highly significant. This suggests that different career facts, including institutional affiliations and networks, are linked to different perspectives or professional lenses through which aquaculture is viewed. Next, we attempt to clarify the composition of these lenses by looking at values.

Values

The role of values in science and expertise is contested. As discussed in Chapter 3, scientists are committed to the notions of value-freedom and value-neutrality as cornerstones of their profession. We attempt to side-step this controversy by making a distinction between "epistemic values" and "non-epistemic values." This distinction is commonly made in the philoso- phy and sociology of science to distinguish between "internal" and "con- textual" values (e.g., Kuhn 1977; McMullin 1983; Rooney 1992). Epistemic values are considered internal to science, and involve commitments to things such as honesty, accuracy, and rigour of study. These are the "governing values and constraints ... [that are] generated from an understanding of what counts as good explanation" within the scientific and expert community (Longino 1990, 4). Non-epistemic values are also referred to as contextual values. These values are the "social, ethical, and political" background to all persons' perspectives on knowledge (cf. Longino 1990).

Our survey focused on *non-epistemic or contextual* values for several reasons. First, as mentioned at the beginning of this book, it was not our intention to evaluate the veracity of competing claims about aquaculture. Second, the survey format is not well suited to testing epistemic values. Scientists and other experts would rarely self-report deviations from honesty, accuracy, and rigour, so measuring adherence to these principles would require ex- tended ethnographic fieldwork similar to the case studies done by Latour and Woolgar (1979) and Knorr Cetina (1999). Third, we expected that non- epistemic values were most important in shaping experts' perspectives on aquaculture. To address such values, the survey included questions on the following value sets: (1) Authority of Science, (2) Principles of Precaution, and (3) Generalized Trust.

Authority of Science
The first set of non-epistemic values was labelled "Authority of Science." We hypothesized that experts with differing stances on aquaculture would hold different views on the capabilities of science and technology to solve prob- lems and understand the machinations of complex environmental systems. We also expected that experts would differ on the role of scientific knowledge in political decision making. The statements used to measure Authority of Science are shown in Table 4.5. Respondents were asked to indicate their agreement with each statement on a five-point scale (indicating strong disagreement, disagreement, neutrality, agreement, or strong agreement). Responses were then coded so that 0 indicates neutrality, −2 indicates strong disagreement, and +2 indicates strong agreement.

To determine whether this value set is associated with different perspec- tives on aquaculture, we performed a correlation test between these responses and the Stance Scale. Correlation tests measure the tendency of variables to vary together. If an increase in one variable can perfectly predict an increase in another, then these variables correlate to 1.0 (or 100 percent). Correlations also measure inverse relationships, where an increase in one variable is as- sociated with a decrease in another (a perfect inverse relationship is scored

Table 4.5

Authority of Science: statements correlated with survey respondents' stance on aquaculture

Statement	Correlation with stance on aquaculture
1 Science and technology can eventually find solutions to most environmental problems.	.405**
2 Science and technology often create as many problems as they solve.	−.251**
3 Scientific knowledge ought to be given more weight than local knowledge in the formulation of environmental policy.	.196**
4 Nature is so complex that it is not fully knowable through scientific investigation.	.135*

Notes: Strong disagreement with each statement = −2; strong agreement = +2.

* Significant at $p < .05$

** Significant at $p < .01$

at −1.0). Table 4.5 shows the correlations between responses to each statement and the Stance Scale. The statements "Science and technology can eventually find solutions to most environmental problems" and "Scientific knowledge ought to be given more weight than local knowledge in the formulation of environmental policy" are positively correlated with stance. This means that the more support for aquaculture an expert professes, the more he or she tends to agree with these statements (and vice versa). The negative correlations show a similar pattern. Here, support for aquaculture is associated with disagreement with the statements "Science and technology often create as many problems as they solve" and "Nature is so complex that it is not fully knowable through scientific investigation." All of the correlations are statistically significant, which means that the relationships are unlikely to be the product of random variation.

The correlations in Table 4.5 are important for understanding the nature of the conflicts among experts on aquaculture. They suggest that experts with differing views of aquaculture tend to hold differing perspectives on the trajectory of scientific and technological development, as well as on the role of science in political decision making. The extent of these differences is not immediately obvious from the data, however, because correlations measure only relative associations, not substantive ones. For instance, we know that opponents of aquaculture tend to *disagree more* with assertions of confidence in the forward march of technological progress, but we don't know whether they *disagree in substance;* they may simply be less enthusiastic. To address this, we must separate respondents into discrete categories.

Table 4.6

Authority of Science: mean scores of survey respondents by stance on aquaculture

Statement	Supporters	Moderates	Opponents
1 Science and technology can eventually find solutions to most environmental problems.[a]	+0.9	+0.5	−0.5
2 Science and technology often create as many problems as they solve.[b]	−0.2	+0.2	+0.4
3 Scientific knowledge ought to be given more weight than local knowledge in the formulation of environmental policy.[c]	+0.6	+0.5	+0.1
4 Nature is so complex that it is not fully knowable through scientific investigation.[d]	+0.5	+0.5	+0.8

Notes: Strong disagreement with each statement = −2; strong agreement = +2.

a $F = 28.3$; significant at $p < .01$
b $F = 5.5$; significant at $p < .01$
c $F = 3.7$; significant $p < .05$
d $F = 1.3$; not significant

In Table 4.6, we consider experts who scored from −14 to −5 on the Stance Scale to be opponents of aquaculture, those who scored −4 to +4 to be moderates, and those who scored +5 to +14 to be supporters. Using these categories, we can compare mean scores on the four statements used to measure views on Authority of Science. Here, we find only two instances of outright disagreement with any of the statements. Opponents of aquaculture disagree with statement 1's assertion that science and technology can eventually solve "most environmental problems," and supporters disagree with the statement that science and technology "often create as many problems as they solve." All other statements are characterized by different levels of agreement.

This finding complicates the discussion, as the data contain several apparent contradictions. For instance, many experts agree with both statements 1 and 2. This may reflect the heightened awareness among all scientists and experts regarding popular and academic critiques of science (as discussed in Chapter 3). Moreover, experts generally agree with statement 4 regarding the limits of science in understanding nature – enough so that the differences in mean score are not statistically significant.[14] At the same time, however, experts on all sides of the controversy agree that scientific knowledge ought to be considered more important than local knowledge in the formulation of policy. Experts are very aware of critiques of science, but continue to advocate its primacy in policy and decision making (we will return to this point later). Thus, despite the convincing correlations shown in Table 4.5, no group of experts ought to be considered highly critical of science. There are clear differences among aquaculture supporters, moderates, and opponents with regard to the four statements, but these are mostly differences of degree rather than substance.

Principles of Precaution
The second value set addressed in our survey concerned the relationship of economic growth to environmental protection. Again using five-point agree/disagree statements, we sought to measure experts' adherence to principles of precaution. The statements are loosely based on the precautionary principle of risk management and environmental protection. The precautionary principle generally asserts that in the absence of scientific certainty or consensus, certain industrial activities and other risky initiatives or behaviours should be severely curtailed or avoided (cf. O'Riordan and Cameron 1994; Harremoes 2002). More specifically, Martin (1997, 266) argues that "the definition, understanding, and application of a precautionary principle hinges on five generic issues, namely likelihood, prevention, danger and/or cost, uncertainty, and possible absence of scientific proof."

Table 4.7 shows the statements that we used to measure Principles of Precaution. It is evident that we did not touch on all of Martin's generic issues, in part because we were concerned about potential overlap with the Stance Scale. For instance, if we were to ask "Is current scientific knowledge capable of addressing the impact of aquaculture over the next 20 years?" we would be addressing an issue rather than a value, and would risk conflating our measures of precaution with those of expert opinion on aquaculture.

Table 4.7

Principles of Precaution: statements correlated with survey respondents' stance on aquaculture

Statement	Correlation with stance on aquaculture
1 Environmental protection should be given priority over economic growth.	−.462*
2 No risk to the environment should be considered acceptable.	−.324*
3 Planning for aquaculture should operate under the assumption that nature has "rights."	−.293*

Notes: Strong disagreement with each statement = −2; strong agreement = +2.
* Significant at $p < .01$

Table 4.8

Principles of Precaution: mean scores of survey respondents by stance on aquaculture

Statement	Supporters	Moderates	Opponents
1 Environmental protection should be given priority over economic growth.[a]	+0.2	+0.9	+1.2
2 No risk to the environment should be considered acceptable.[b]	−0.8	−0.3	0
3 Planning for aquaculture should operate under the assumption that nature has "rights."[c]	+0.2	+0.5	+1.1

Notes: Strong disagreement with each statement = −2; strong agreement = +2.

a $F = 28.1$; significant at $p < .01$

b $F = 10.4$; significant at $p < .01$

c $F = 9.4$; significant at $p < .01$

To avoid this, we can only address those elements of precaution that *precede* a specific issue. We are therefore limited to examining "danger and/or cost" as well as questions of priorities (which we would add to Martin's list).

As Table 4.7 shows, the correlations between each item and stance on aquaculture are again strong. Each case demonstrates a negative correlation, so that stronger agreement with the statements is associated with stronger criticism of aquaculture. Table 4.8 reports the mean scores for these statements according to the grouped stance on aquaculture categories discussed previously.

As with the Authority of Science items, the strong correlations mask some marked instances of agreement. Experts on all sides agree with statements 1 and 3, and nearly all disagree with statement 2 (with opponents scoring an exactly neutral "0"). This agreement-in-substance regarding the balance of economic and environmental priorities is notable. Nevertheless, the relative strength of adherence to precautionary values is a strong predictor of expert opinion on aquaculture. This is a substantial component of the values lens through which experts view aquaculture.

Generalized Trust

The values measured by the Authority of Science and Principles of Precaution items are important because they reflect the approach that experts adopt when addressing professional questions and issues. Basic or generalized trust is also an important component of the values context in which experts interpret aquaculture. Trust is a fundamental social attribute. It is, in Yamagishi's words (2001), "a form of social intelligence": a foundation for evaluating the world and conducting oneself within it. According to authors such as Putnam (1993) and Stolle (1998), the most significant form of trust where social relations are concerned is generalized trust. Generalized trust in others is essentially trust without proofs – it is the benefit of the doubt that is given to "unfamiliars," to strangers and/or strange circumstances. Thus, it has been argued that generalized trust is a pillar of participatory democracy (see Putnam 1993) because, even when it does not produce consensus, generalized trust reduces the potential for conflict by promoting an openness to other viewpoints and actions. According to Luhmann (1979, 24-25), trust is a key means of "reducing social complexity":

> The world is ... dissipated into an uncontrollable complexity; so much so that at any given time people are able to choose freely between very different actions. Nevertheless, I have to act here and now. There is only a brief moment of time in which it is possible for me to see what others do, and consciously adapt myself to it. In just that moment only a little complexity can be envisaged and processed, thus only a little gain in rationality is possible ... Trust, by the reduction of complexity, discloses possibilities for action which would have otherwise remained improbable and unattractive without trust.

According to this perspective, generalized trust both filters information (complexity) and fills in information (lack of direct knowledge about other actors). For these reasons, generalized trust is here considered a key component of the lens through which experts view aquaculture and its attendant controversy. Generalized trust is an important guide for experts who must evaluate competing claims and complex relationships in the context of controversy.

The survey contained several questions regarding trust. Most of these, however, refer to different groups associated with the aquaculture controversy, and we will turn to them in the next section. Our measure of generalized trust involved asking respondents to indicate level of disagreement or agreement with the statement "Most people can be trusted." This simple statement is the most common measure of generalized trust and has proven to be valid over a range of studies (e.g., Grieder et al. 1991; Veenstra 2000; Putnam 2000). Using this measure, we find that generalized trust and stance on aquaculture correlate to .24. The positive correlation indicates that experts who agree *more* with the generalized trust statement tend to be *more supportive* of aquaculture. Conversely, lower levels of generalized trust are associated with less supportive stances on aquaculture.

This is an important finding, as it suggests that opponents of aquaculture tend to demonstrate more general skepticism than moderates or supporters. That is, skepticism of the aquaculture industry and of "generalized others" appear to be linked, although we ought to consider other possibilities. For instance, lower general trust may be a product of involvement in activism, which is by nature somewhat confrontational. The substantive data on trust show that no category of respondents are active distrusters, however. Supporters of aquaculture achieved a mean of +0.6, moderates a mean of +0.4, and opponents a neutral 0.[15]

Although modest, these differences in trust ought to be considered as part of the lenses that lead experts to differing interpretations of the effects of aquaculture. If trust (and mistrust) are indeed means of reducing social complexity, our findings suggest a link between this and the interpretation of scientific and other data in the formulation of expert opinion.

Drawing Tentative Conclusions about Experts' Stance on Aquaculture
As mentioned at the outset, in the course of our research we had many informal conversations with aquaculture experts on all sides of the controversy. These discussions showed us that although deep disagreements exist, many experts hold nuanced and complex views of the industry. The survey gives us a means to investigate these disagreements, and to look for patterns that might lend insight into the strength and duration of the controversy. In line with this, we have seen that there is some common ground among experts regarding positive economic outcomes for coastal communities (in Chapter 2, we saw that this question is fiercely contested at the local level). In contrast, issues of environment and human health are deeply divisive. With respect to overall assessments or stance regarding aquaculture, we have seen

Table 4.9

Percentage of variance on the Stance Scale explained by institutional affiliation, individual situation, and values, based on a three-step regression model

Step	Variables considered	Percentage of total variance explained
1	Institutional affiliation	16
2	Institutional affiliation + individual situation	28
3	Institutional affiliation + individual situation + values	46

Notes:
1 The final regression equation for Step 3 is: $y = -3.44$ (university affiliation) $- 6.43$ (ENGO affiliation) $+ 5.71$ (university or college affiliation; previous employment in industry) $+ 1.04$ (industry affiliation; collaboration with industry colleagues) $+ 1.24$ (statement 1, Authority of Science) $+ 1.01$ (statement 3, Authority of Science) $- 1.90$ (statement 1, Principles of Precaution) $- 0.88$ (statement 2, Principles of Precaution) $+ 0.95$ (Generalized Trust) $+ 17.17$ (constant).
2 Each variable is significant at $p < .01$, except "industry affiliation; collaboration with industry colleagues" and "generalized trust," which are significant at $p < .05$.

Table 4.10

___

Percentage of variance on the Stance Scale within each institutional category explained by individual situation and values

Institutional category	Percentage of variance explained
University or college	42
Government	34
Industry	55
Environmentalist group	–

___

that institutional affiliation, professional network, and contextual (non-epistemic) values each strongly relates to experts' overall understandings of aquaculture. In statistical terms, each of these partially explains the overall variance on the Stance Scale, and thus in part predicts an individual respondent's score.

The overall power of our measures to predict experts' stance on aquaculture is shown in Table 4.9. This table presents the total amount of variance in experts' stances on aquaculture that can be explained at each step, where we introduce new variables into the regression model. The end result is 46 percent explained variance, which is high given the diversity of opinions and assessments of aquaculture revealed in Table 4.2.[16]

Finally, Table 4.10 demonstrates the amount of variance explained within the institutional categories of academia, government, and industry. These high numbers show that individual situation and values are also applicable within the different institutional settings. The high proportion of explained variance for experts affiliated with industry is particularly noteworthy, as it suggests that even in this setting, where opinions are generally consistent (see Table 4.2), differences in values and career activities are strongly associated with measurable differences in experts' views of aquaculture.

Ultimately, the patterns that we find prompt us to argue that experts are interpreting conflicting claims about aquaculture from substantively different vantage points, or through lenses that are strongly related to their professional experiences and personal values. This suggests that disagreements about aquaculture are more than just about facts, but are influenced by the flows of information and interpretation through peer groups and by the contextual values that guide experts' broader views of the world.

Experts' Views of One Another
Experts' views of one another are fundamental to the future of the aquaculture controversy. Scientific consensus on a given question tends to emerge incrementally, as hypotheses are advanced and modified by a community

of peers (Kuhn 1970). Although new trails are often blazed by mavericks or radically original thinkers, consensus is produced by the careful and sometimes plodding work of scientific communities. This process relies critically on mutual trust; scientists and other experts need to trust the findings and claims of others in order to build on them (Hardwig 1991), and protracted struggles over knowledge claims tend to erode trust (cf. Slovic 1999). This section therefore concerns one of the most serious consequences of the conflict over aquaculture – the widespread mistrust and hostility within the community of aquaculture experts. The high-profile activism and conflict between environmentalist groups and industry supporters that we examined in Chapter 3 implicate only a small minority of aquaculture experts, but we suspect that these conflicts have a poisonous effect across the expert population, a suspicion that our data, unfortunately, confirm.

In this section we also go beyond the statistics to look at how experts *explain* problems with trust. Using data from open-ended questions in the survey, we find that many experts advance an alternative explanation for mistrust that vindicates scientists themselves and highlights the abuse of science by "outsiders" – particularly government, industry, and environmentalist groups. We interpret this as an argument in defence of the integrity of science and expertise, and as a statement that consensus is not out of reach if only the scientific community could be freed from outside influence.

The Question of Trust

Experts' views of one another are not easy to measure. Disagreements among aquaculture experts are often cordial and professional, but they can also be personal and, to put it delicately, unscientific. Our survey provided an opportunity to measure the consequences of these disagreements, but in an ethical way. For this reason, we designed the survey to avoid references to specific people. Instead, our measures of experts' views of one another are based only on broad categories. In the previous section, we saw that there is a correlation between generalized trust and stance on aquaculture, with opponents of aquaculture tending to be less trusting in a generalized sense; we speculated that this might be linked to experiences with activism. Here, we look for patterns in specific levels of trust and mistrust – that is, the judgments that experts have of others involved in the aquaculture debate. Our main measure for this is the question "How trustworthy are the following to produce good science about aquaculture?" Respondents were asked to indicate whether they judged scientists affiliated with universities, scientists affiliated with government, scientists affiliated with the aquaculture industry, and scientists affiliated with environmentalist groups as being very untrustworthy, somewhat untrustworthy, neutral, somewhat trustworthy, or very trustworthy (yielding a scale of -2 to $+2$).

Table 4.11

Correlations between survey respondents' trust in different groups of experts and the Stance Scale

	Trust in scientists			
	In academia	In government	Affiliated with industry	Affiliated with environmentalist groups
Correlation to Stance Scale	−.02	.286*	.493*	−.325*

Notes: For all groups, survey responses ranged from −2 (very untrustworthy) to +2 (very trustworthy).
* Significant at $p < .001$

Table 4.12

Trust in different groups of experts: mean scores of survey respondents by stance on aquaculture

	Trust in scientists			
Stance on aquaculture	In academia	In government	Affiliated with industry	Affiliated with environmentalist groups
Opponent	+1.1	+0.2	−1.1	0
Moderate	+1.1	+1.0	+0.3	0
Supporter	+1.1	+1.1	+0.6	−0.7

Notes:
1 For all groups, survey responses ranged from −2 (very untrustworthy) to +2 (very trustworthy).
2 Analyses of variance on this measure are significant at $p < .01$ with the exception of trust in scientists affiliated with academia (not significant) (*F*-test).

Because trust plays such an important role in science, we would normally expect most evaluations to be on the high end of the scale (i.e., towards evaluations of +2, or "very trustworthy") (Hardwig 1991). This is not reflected in the data, however. Table 4.11 shows that levels of trust in other experts correlate strongly with the Stance Scale that we discussed earlier. For instance, trust in government-affiliated scientists and industry-affiliated scientists tends to increase among respondents who hold a more positive view of aquaculture. The reverse is true of trust in scientists affiliated with environmentalist organizations, where a more negative assessment of aquaculture is correlated with higher levels of trust in that group of experts. This suggests that experts' views of one another generally align with their stance on the issue itself, that professional agreements and disagreements are linked with levels of trust in the competence of others.

Again, however, there is more to the story. Table 4.12 presents the mean scores on this question across supporters, moderates, and opponents with regard to aquaculture. It is immediately evident that university-based scientists are the most trusted group, and are equally trusted across all sides of the debate. As discussed earlier, this is probably in part because universities are diverse institutions, housing all types of aquaculture experts. It is also probably in part due to the prestige of academia, specifically the notion (true or not) that academia is a politically disinterested pursuit (McCollow 1996). There is significant variance in levels of trust across the other institutional categories. For instance, we see that government-affiliated scientists are more trusted by supporters of aquaculture than by opponents, and a similar pattern is found for trust in industry-affiliated scientists. Overall, the lowest trust scores are among opponents of aquaculture towards industry-affiliated scientists, and among supporters of aquaculture towards scientists affiliated with environmentalist groups. Interestingly, environmentalist groups are the least trusted of any institutional category. This finding contradicts the usual discovery that environmentalist groups are highly trusted in controversies, at least by members of the general public (Yearley 2003, 41). According to our survey, even opponents of aquaculture are less trusting of scientists affiliated with environmentalist groups than they are of government- or university-based scientists.

These latter findings are also important because they suggest that experts are not necessarily closing ranks on questions of trust. The generally poor evaluation given by opponents of aquaculture to the science produced by environmentalist organizations may reflect a professional discomfort about the controversy and some of the more extreme claims and statements made in activist publications (see Chapter 3). Indeed, it is also interesting to note that respondents who are themselves affiliated with industry or environmentalist groups ranked their own institutions as less trustworthy than academia. For instance, experts affiliated with environmentalist groups on average evaluated university-based scientists higher (+1.3) than their own groups (+0.8). Similarly, experts affiliated with industry considered university-based scientists to be slightly more trustworthy than industry-based scientists (+0.9 versus +0.8).

Dilemmas of Trust and Science

Qualitative data from the survey enable us to take a more careful look at issues of trust and mistrust. Our findings appear contradictory at first. On the one hand, we see that accusations of bias are commonplace. Labelling researchers or claims as biased demonstrates severe mistrust, because it implies a purposeful dishonesty. It suggests that science has taken a backseat to political considerations, which would be a betrayal of the fundamental ideology of science, particularly the principles of value-neutrality and community.

On the other hand, we see that many experts have an alternative explanation for problems of mistrust. For them, the problem lies in the application of science by others. This is a key distinction, as it permits a continued faith in the community of experts on aquaculture by assigning blame to others who are either outside or at the margins of science.

We first consider accusations of bias, which are often expressed in broad or blanket statements about adversaries. Following the closed questions on trustworthiness, the survey invited participants to comment under the heading "We would like to give you the opportunity to comment further on issues of trust." Responses included:

The level of trust is very bad. People are behaving in untrustworthy ways, for the bottom line or to protect their jobs. There are no principles or vision being shown anywhere. Everyone seems to have an agenda. (Respondent 21: Unknown Affiliation; Oppositional Stance)

Environmental groups should be held responsible for their science to the same level industry and government scientists are (by media and peer reviews). Hypotheses should be identified as such, and not as "gospel." (Respondent 22: Industry Affiliation; Supportive Stance)

A major problem is industry scientists['] frequent refusal to accept negative findings (assuming these have been scientifically verified and are valid) ... Most frequently, [they] persist in a state of denial. (Respondent 139: Government Affiliation; Moderate Stance)

Such sentiments are common in the open-ended data. Roughly one-half of respondents took the opportunity to comment on questions of trust, and roughly one-half of responses contained a negative evaluation of others. As with the foregoing quotations, some comments referred specifically to scientists and their behaviours. In other cases, auxiliary agencies such as governments and funders are invoked in explanations of imbalanced, "bad," or biased science:

Governments are ignoring the sound scientific data and facts that are actually out there. As such they are encouraging the widespread misperceived notions, propaganda, and misinterpreted science that are put out there by bad scientists. (Respondent 50: University Affiliation; Moderate Stance)

[We need to] develop sustainable aquaculture practices [that are] based on real scientific data. We cannot have science mandated by public opinion polls or multinational environmental NGOs. (Respondent 14: Industry Affiliation; Supportive Stance)

The narratives about "bad science" and "bad scientists" suggest that mistrust among aquaculture experts can be extreme. Trust or mistrust are not simply present or absent – they are phenomena of degree (Lewis and Weigert 1985, 975). For instance, skepticism can be interpreted as a form of mistrust, but one that is mild or even beneficial ("a healthy skepticism"). In contrast, accusations of bad science that are levelled by all sides in the debate are significantly more troubling, suggesting that the controversy is poisoning experts' views of one another.

On the other hand, the open-ended commentary also revealed an important alternative explanation for bias in science. According to some aquaculture experts, bias comes from outside actors:

> Scientists create knowledge – it is in the application of this knowledge that the problems with trust occur. Usually someone other than the scientist is responsible for this [as well as for poor] interpretation of data. (Respondent 13: University Affiliation; Supportive Stance)

This explanation is important because it involves a clear statement of the boundaries between scientists and users of science, largely absolving the former and faulting the latter (Young and Matthews 2003):

> All the scientific advice in the world can be provided to politicians and policy makers who can choose to ignore advice to meet their objectives, such as the underlying need for re-election and payback of campaign promises. (Respondent 70: Aboriginal Affiliation; Oppositional Stance)

> I think that the bad records [on trust] come more from what institutions do to spoil scientists' good intentions. For example, the government uses their scientists badly, so [the scientists] look bad ... To insure IMPROVED and honest reporting and to counter foolish statements made by organizations with set agendas, industry and government scientists must do a much better job in communicating fact and truth. (Respondent 260: University Affiliation; Supportive Stance)

The outsiders-as-biased argument recasts the question of trust. Whereas the earlier quotations clearly demonstrate active mistrust among experts, this narratives leaves the door open. The issue of mistrust is not resolved, but it is displaced in a way that protects the integrity of science as an honest pursuit that may *eventually* lead to resolution. This can lead to some very progressive and thoughtful propositions for the resolution of the problem of trust:

> Trust is key. Unfortunately, the "debate" around aquaculture has evolved to such a point where it becomes a matter of being "for" or "against." Somehow

aquaculture seems to generate the same response as abortion or capital punishment ... Trust is central to these issues because folks seem to have become sceptical of industry, academia, government and ENGOs. Everybody needs to open their institutions to public scrutiny in order to begin building trust. (Respondent 253: Government Affiliation; Supportive Stance)

Anyone who is out to promote something or to be "right" cannot be trusted. Until there is a joint, collaborative effort among First Nations [i.e., indigenous peoples], industry, government and environmental groups to agree on methodology and conduct unbiased science, science will continue to get bogged down in issues of who is right and who is wrong. (Respondent 246: Government Affiliation; Oppositional Stance)

So, what conclusions can we draw from these findings about the problem of trust? First, the data show that trust is patterned. Experts are willing to trust the science produced by certain institutions more than that produced by others, and this willingness is strongly associated with an expert's stance on aquaculture (with the notable exception of trust in academia, which is universally high). Furthermore, the data show that mistrust is strongest at the extremes. For example, opponents of aquaculture are very mistrusting of the capacity of industry-affiliated scientists to produce good science about aquaculture, while supporters of aquaculture are even more suspicious of scientists affiliated with environmentalist groups. Much of the open-ended commentary given by respondents reinforces this conclusion: mistrust is endemic, and it leads to very critical assessments of the motivations and behaviours of others.

At the same time, however, some experts have adopted an alternative explanation of the trust problem that displaces blame. These explanations suggest that the problem occurs *outside* of science – in the interpretation and application of science by non-scientists or non-experts. This interpretation protects the integrity of science by distancing it from the politics that occur elsewhere. Moreover, it allows for the possibility of resolving the trust problem by, in the words of one respondent, making sure that "scientists do a much better job in communicating fact and truth," which indicates a strong faith in the community of aquaculture experts despite the controversy. This explanation comes even more strongly into play when experts are asked about their views of the public.

Experts' Views of the Public
For a long time, it was assumed that experts had a one-way relationship with the general public (see Chapter 3). Experts were experts because they held special knowledge, and this knowledge was assumed to be above politics and removed from the messy give-and-take of public and political debates

(Fischer 2000). This view is less and less tenable. We saw in Chapter 2 that stakeholder claims about aquaculture are forceful and complex, often alternately using and challenging scientific claims. Looking specifically at British Columbia, it is instructive to note that the first official review of the aquaculture industry, the Gillespie report in 1986, was conducted in consultation with experts only. Since then, periodic government reviews have included and embraced public input. The Salmon Aquaculture Review, completed in 1997, involved consultations with coastal First Nations, leading to a dedicated volume of the final report entitled *First Nations Perspectives* (SAR 1997b). More recently, the Special Committee on Sustainable Aquaculture in British Columbia conducted dozens of open meetings with coastal First Nations and non-Aboriginal communities in 2006 and 2007 to go along with expert reviews of aquaculture regulations and economic impacts (SCSA 2007a, 2007b). In short, although questions of science dominate the aquaculture controversy in Canada, these questions are no longer the exclusive province of experts. Stakeholders and the general public are elbowing their way onto the knowledge battlefield.

The rising influence of stakeholders and the general public is something of a wildcard in scientific and environmental controversies. First, stakeholder and public sentiment is often difficult to predict, and can turn quickly following a gaffe or misstep (Leiss 2001). Second, it is often unclear how scientists and other experts will react to public involvement in scientific issues. According to some research, experts are often reluctant to meaningfully engage with lay actors, seeing consultation as an opportunity to educate rather than involve stakeholders (cf. Wynne 2001; Hannigan 2006). On the other hand, some research suggests that experts are genuinely open to stakeholder participation in formulating and evaluating expert knowledge, and that this goodwill is notable given the implicit challenge that this participation poses to the privileged status accorded to traditional forms of expertise (Young and Matthews 2007a; Blok 2007).

In this section, we again use findings from our survey to look at experts' views of public participation in the aquaculture controversy.[17] As we have seen, the expert community is deeply divided in its assessments of aquaculture, and is troubled by mistrust and hostility. This leads us to develop two sets of hypotheses regarding experts' views of public participation in the controversy.

First, we expect opponents of aquaculture to be more open than supporters to direct public involvement. Opponents are generally critical of the industry and of government sponsorship and regulation of aquaculture, and are likely to see public involvement as a means of furthering this critical agenda. Following this logic, we expect supporters of aquaculture to be wary of greater public participation following years of conflict with environmentalist and other opposition groups. Our second hypothesis, however, is that experts on all sides of the debate will have complex views of the public. Specifically, we expect these experts to recognize that public participation is necessary and may lead to positive outcomes; at the same time, however, we expect them to be somewhat suspicious of stakeholder and public involvement. As we have seen, the aquaculture controversy is trust eroding, and we expect this to apply also to experts' views of the public.

In what follows, we will look at three dimensions of experts' views of the public. First, we will examine their views on stakeholder participation in aquaculture. We have seen that aquaculture is highly controversial at the local level, and that both pro- and anti-aquaculture groups are working directly with stakeholder groups to advance their agendas. Second, we will look at experts' views of the media. Although we will consider in detail the content of media coverage of aquaculture in Chapter 5, experts' views of the media are an important dimension of their understanding of public participation in the controversy. The media are an important bridge between experts and the public. They are also a "public actor," in the sense that they play a role in steering public debates and amplifying certain issues and narratives (Kasperson et al. 2001). Third, we will look at experts' views of the general public, specifically the knowledge and values basis of general public opinion. Overall, we find that experts' views of the public swing from strongly positive to strongly negative across these dimensions.

Experts' Views on Stakeholder Participation

Our survey contained two closed-ended questions regarding stakeholder participation in aquaculture: (1) "Local knowledge ought to be included in formal risk assessments," and (2) "Aboriginal knowledge and concerns should be reflected in policies dealing with aquaculture." As with the statements discussed in previous sections, respondents were asked to indicate whether they strongly disagreed, disagreed, were neutral, agreed, or strongly agreed with these statements, yielding a score from –2 to +2. Using these scores, we can correlate responses with experts' opinions on aquaculture (the Stance Scale). On the "local knowledge" statement, the correlation is a modest but significant –.22, meaning that experts who are more supportive of aquaculture tend to be *less* enthusiastic about the inclusion of local knowledge in formal risk assessments. Similarly, the "Aboriginal concerns" statement correlates with stance to –.33. These findings support our first hypothesis – that opponents of aquaculture are more open to stakeholder participation in formal aquaculture practices and policies.

Table 4.13, however, gives the mean scores achieved by respondents according to the discrete categories that we constructed based on the Stance Scale (aquaculture opponents, moderates, and supporters), as well as by institutional affiliation (university or college, government, industry, or environmentalist group). This table shows that all categories of experts *agree*

Table 4.13

Views on stakeholder participation: mean scores of survey respondents by stance on aquaculture and institutional affiliation

	"Local knowledge ought to be included in formal risk assessments"	"Aboriginal knowledge and concerns should be reflected in policies dealing with aquaculture"
Stance on aquaculture		
Opponent	+1.26	+1.29
Moderate	+1.13	+0.88
Supporter	+1.00	+0.67
Institutional affiliation		
University or college	+1.19	+0.91
Government	+1.03	+0.74
Industry	+0.84	+0.60
Environmentalist group	+1.19	+1.50

Notes:

1 For both statements, survey responses ranged from –2 (strongly disagree) to +2 (strongly agree).

2 Differences between the categories based on stance on aquaculture and on institutional affiliation on both statements are significant at $p < .01$ (*F*-test).

with both statements. It is important to note that although levels of agreement scale up from supporters of aquaculture to moderates to opponents on both items (hence the correlations), these differences are *only* in degree of agreement.

The general agreement with these statements is important because it (in part) contradicts our first hypothesis: opponents of aquaculture may be more open to stakeholder participation, but supporters are open to it as well. There are several probable explanations for this. First, we have to assume that agreement with the statements is politic, particularly with regard to Aboriginal issues (although, intriguingly, this statement receives less endorsement than the local knowledge statement). Second, the text of the statements ought to be critically considered (and defended). To begin with, both are cast positively and both contain normative statements ("ought to" and "should be"), which may skew the responses. Also, the text is ambiguous despite referring to the specific processes of "formal risk assessments" and "policies dealing with aquaculture." We justify the text of the statements by stressing the challenge of drafting topical statements about potential stakeholder participation without referring to any side of the controversy, which would risk conflating the statements with those used in the Stance Scale (see Figure 4.1). For instance, the survey also contains the statement "Aboriginal groups should be given the right to veto the siting of aquaculture operations."

Clearly, this statement risks interacting with Stance, as the reference to "veto" might encourage supporters and opponents of aquaculture to line up on either side. Thus, our statements sacrifice some precision in an effort to strike a middle ground that is not inherently contrary to any position or stance in the debate.

The third explanation is the most important, namely, that experts on all sides of the debate have a genuine desire to engage with stakeholders. We find support for this in the open-ended commentary discussed below.[18] At the same time, however, we also see that many experts, again on all sides of the debate, want public participation to be limited rather than fundamental.

To begin, the commentary given by respondents reveals that many experts hold nuanced views on the relationship between different types of expert and lay knowledge:

Values are implicit in any knowledge system. I think that this needs to be taken into account in evaluations and in the use of knowledge. Risk increases with the competitive exclusion of knowledge types. The simplest solutions for dealing with risk come from people who make risk prone decisions (and their families) [who must] live with the consequences of their decisions. (Respondent 15: Aboriginal Affiliation; Moderate Stance)

TEK [Traditional Ecological Knowledge] ... exists, [and] it is not restricted to Aboriginal people. It comes from direct experience and contact with nature and "elders" who have learned from experience. There must be continuous contact with the environment for TEK to flourish. Scientists should take the observations and experiences of "TEKers" and ask what is the underlying science rather than dismiss it right out of hand. However, too much weight is also given to TEK by some politicians and their advisers and it is taken out of hand. (Respondent 122: University Affiliation; Supportive Stance)

These comments sincerely recognize the substantive and unique contributions of local or stakeholder knowledge to the understanding and management of the effects of aquaculture (cf. Fiorino 1990, 227). The open-ended commentary also contains some themes and trends, however, that give insight into how many respondents understand the notion of stakeholder participation in somewhat narrow terms. For example, a widespread trend involves clearly establishing the priority of scientific knowledge over local knowledge or concerns. This view is particularly prevalent among moderates and supporters of aquaculture, but is also expressed by opponents.

First and foremost, unbiased scientific information is critical in determining what practices are appropriate, in what locations, etc. This goes hand-in-hand with environmental concerns because aquaculture and the surrounding environment undoubtedly interact with each other. The general community and specifically the Aboriginal community *need to be considered next* [emphasis added]. (Respondent 164: Government Affiliation; Supportive Stance)

Scientific data ... go hand-in-hand with environmental issues. Ideally, industry's objectives would be supported by communities who hope to benefit from aquaculture, but all too often it seems decisions are guided by lobby rather than hard facts. (Respondent 207: Government Affiliation; Moderate Stance)

[Aboriginal people] should be involved when their communities are involved. However, they cannot go into the discussion without good training and experience in aquaculture. (Respondent 56: University Affiliation; Supportive Stance)

Government policy should stand on a balance of ... environmental and community concerns and desires [but] *based on REAL* scientific data and results [emphasis added]. (Respondent 285: University Affiliation; Moderate Stance)

We CAN answer the questions about likely risks from aquaculture ... The only way to know is to ask the questions and collect DATA to answer them [and] NOT opinions, which seems to be the current approach. (Respondent 201: University Affiliation; Oppositional Stance)

Each of these comments draws some form of distinction between scientific and non-scientific claims. This adds some texture to the willingness of experts on all sides of the debate to include "local knowledge" and "Aboriginal knowledge and concerns" in formal assessments and regulatory activities. The open-ended responses generally demonstrate that experts have a deep awareness of the pressures to address and incorporate lay claims and concerns into instruments of research and policy, but they also demonstrate that for many experts, science remains the primary arbiter of contentious issues, with local knowledge taking on a complementary role. This is even more evident in the following comments, which envision specific roles for community or local people with respect to knowledge:

[Aboriginal people's] concerns should be addressed, and their rights respected in siting and operating aquaculture sites ... I would prefer to see their role as participating in audit or review committees as opposed to conducting separate monitoring efforts. They could have input to monitoring requirements that are being carried out by industry or government through such an overview committee. (Respondent 140: Government Affiliation; Moderate Stance)

There is a need for more cooperation on all sides. Industry should be funding projects that are led by universities, but where government (scientists) and community ... groups can participate in. This is an open transparent method where everyone is involved and it shows industry is making an effort to consider their impacts to the environment and that they are not just concerned about the economics. (Respondent 210: Government Affiliation; Moderate Stance)

These two comments demonstrate awareness of public skepticism of black-boxed science, and both propose solutions based on "audit or review" or "open transparency" through participation. In both cases, however, the inclusion of local actors is posited against an alternative. In the first statement, the inclusion of Aboriginal people is "as opposed to conducting separate monitoring efforts." Here, "input" is being substituted for "efforts" towards a parallel process that would stand outside of official science. In the second statement, community participation is invited to show that "industry is making an effort to consider their impacts to the environment." In this case, community participation is equated with transparency and legitimacy, rather than considered as a source of potential challenge and critique of projects funded by industry and led by universities.

To be sure, both comments are laudable and well-meaning, and both reflect the distance that has been travelled towards greater institutional reflexivity in science through awareness of potential criticism of "closed" science (cf. Wynne 1993; Demeritt 2000). They reflect, however, what Fiorino (1990, 228) terms an "instrumentalist argument" for stakeholder engagement. According to this, stakeholder involvement is understood as heightening the legitimacy of expert practices and findings, rather than making a substantive contribution to these. So while experts on all sides of the controversy are markedly receptive and even enthusiastic about local participation in risk assessment and policy, lay participation is framed as complementary to scientific processes, or as enveloped within processes that are led by science. As such, the potential critique levelled by local knowledge on the dominance of scientific knowledge is diminished, as local is something "to be considered next," or else directly enlisted in the interests of transparency.

Experts' Views of the Media
We now shift emphasis to the understanding that experts have of the wider public with respect to the aquaculture controversy. As discussed previously, aquaculture is a significant risk issue in Canada that has garnered high levels of media and public attention. Given the public profile of this controversy,

we expect aquaculture experts to be highly aware of the importance of public opinion in the formation of aquaculture policy. In the following analysis, therefore, we seek a better understanding of two phenomena: how experts view the presentation, interpretation, and use of science in public discourse, and how experts understand their relationship with the wider public.

A significant academic literature exists regarding the relationship between scientists and the media (e.g., Friedman et al. 1999; Kasperson et al. 2001; Cox 2006). In some conceptual models, the media are considered to be a passive vehicle for science to use in speaking to the public. The traditional approach to science communication is described by Burns and colleagues (2003) as a unidirectional system where "specialists known as mediators [and] science communicators" work through the media to present scientific findings to the public. More critical studies examine the tendency of the media to frame scientific claims in exaggerated or moralistic terms, particularly when claims are contested or controversial (e.g., Roll-Hansen 1994; Malone et al. 2000; Schabel 2003). Finally, recent survey evidence suggests that experts and journalists hold very different ideas about the role of the media in science communication. For instance, Peters (1995, 38) argues that differences in the professional cultures of scientists and journalists contribute to different understandings of the purpose of science communication: "journalists are more apt to claim a watchdog function of being critical towards those in power than are experts willing to concede it." Similarly, Hartz and Chappell (1997) find that mutual distrust between scientists and journalists is pervasive, with each group claiming that the other misunderstands the demands and norms of the group's profession.

Our aim here is to explore aquaculture experts' perspectives on the media as a communicator of aquaculture science. As with views on stakeholder participation, we anticipate that an expert's "location" within the controversy (stance) may relate to his or her views on the media. Moreover, our survey indicated that a large proportion of respondents (57 percent) had given at least one interview "related to aquaculture" to the popular media in the preceding five years. Thus, most of the experts whose views are analyzed here have had direct experience with media interpretation and representation of claims or arguments.

Our survey contained two questions asking respondents to indicate level of agreement with the statements: (1) "The media do a good job representing aquaculture science to the public," and (2) "The media can be trusted to portray aquaculture in a responsible manner." These questions were placed in different sections of the survey to avoid overly patterned responses. As anticipated, responses to these statements correlate strongly with stance, to −.38 on the first and −.23 on the second. Thus, experts who are more supportive of aquaculture tend to have poorer opinions of media representation

Table 4.14

Views of the media: mean scores of survey respondents by stance on aquaculture and institutional affiliation

	"The media do a good job representing aquaculture science to the public"	"The media can be trusted to portray aquaculture in a responsible manner"
Stance on aquaculture		
Opponent	−0.63	−0.92
Moderate	−0.68	−0.85
Supporter	−1.28	−1.28
Institutional affiliation		
University or college	−0.85	−0.92
Government	−0.95	−1.03
Industry	−1.48	−1.64
Environmentalist group	−0.36	−0.75

Notes:

1 For both statements, survey responses ranged from −2 (strongly disagree) to +2 (strongly agree).

2 Differences between the categories based on stance on aquaculture and on institutional affiliation on both statements are significant at $p < .01$ (*F*-test).

of aquaculture science, and less trust in the media to portray aquaculture "in a responsible manner."

Although the correlations between views of the media and stance are substantial (as with views on stakeholder participation), they again mask relative consistency. In this case, however, the evaluation is overwhelmingly negative. Table 4.14 gives the mean scores for the two statements, according to stance and according to institutional affiliation. The universally poor assessments of the media are evident, reaching across every category of expert in the aquaculture controversy.

As with the statements addressing stakeholder participation, the statements used to measure views of the media are framed somewhat ambiguously. Again, this was necessary in order to avoid offending experts on any one side of the debate and thus artificially skewing the measure. The consequence of this necessary ambiguity is that we are not exactly certain how experts interpret the phrases "representing aquaculture science" and "portray aquaculture in a responsible manner." In other words, the quantitative data cannot inform us about the roots of these overwhelmingly poor assessments. We again turn to the commentary provided by respondents for elaboration.[19] In the comments regarding the media, the most obvious trend involves conflicting accusations of bias. First are two examples of perceived anti-aquaculture bias:

> The public is influenced by the news media, which is fixated on the need to produce controversy and therefore tends to feature only undesirable aspects of the industry. (Respondent 207: Government Affiliation; Moderate Stance)

> Any knowledge gained by the public seems to come from the media. The media are driven by their markets and owners' political leanings and what they think the public wants to hear. News articles which attack the industry seem to sell and thus, regardless of the validity of the science, get aired. This creates a vicious cycle of misinformation. (Respondent 37: Government Affiliation; Supportive Stance)

In these comments, the media are directly accused of misrepresenting the industry and the true state of affairs to the public. Accusations that the media have a pro-aquaculture bias are equally common, however. In the following comments, the media tend to be accused of *being a vehicle* for the misrepresentation of truth to a public that is vulnerable to misinformation:

> Industry regularly misrepresents the truth to the public [through] their press releases and media spots versus some very conclusive evidence to the contrary on a number of issues. (Respondent 139: Government Affiliation; Moderate Stance)

> [The public only ever] sees the glossed-over view presented by [salmon] farmers' paid spin doctors and not the mess and reality that is fish farming. (Respondent 70: Government Affiliation; Stance unknown)

These conflicting accusations of bias are consistent with a phenomenon that has been termed "the hostile media effect," whereby partisans of a controversial issue typically perceive media coverage as hostile to their position (Vallone et al. 1985; Gunther and Schmitt 2004). Indeed, the thematic consistency of the allegations from all sides of the controversy strongly suggests that "bias" is an easy tag or explanation for a widespread frustration with media portrayal of aquaculture. In Chapter 5, we will look closely at trends in media coverage of aquaculture in Canada. Although there is much to criticize about media reporting on this issue, we find little evidence of systematic bias. Of the 1,500 newspaper items analyzed, three-quarters adopt a point/counterpoint position (where arguments for and against aquaculture are presented together). Among the one-quarter of items that adopt a dominant position (the vast majority of which were editorials and letters to the editor), 56 percent express only opposition to the industry and 44 percent

express only support. Interestingly, our survey found no significant association between contact with the media and views of the media. As mentioned,
57 percent of respondents indicated that they had given at least one media
interview "related to aquaculture" in the preceding five years. There are no
significant differences of opinion on the media between experts who have
been interviewed and those who have not; both are strongly negative.

Recent work by Gunther and Schmitt (2004) on the hostile media effect
offer a partial clarification of our findings. These authors conducted an experiment in which a purposely crafted neutral text was presented to experts
involved in the ongoing controversy over genetically modified organisms.
For one randomized group of experts, this text was presented as a news item;
for the other, the identical text was presented as a research paper from a
senior undergraduate student. In comparing participants' evaluations of bias
in the text, Gunther and Schmitt found striking differences. Whereas the
presentation of the text as a news item yielded extreme and contradictory
assessments of bias, the identical text presented as an undergraduate research
paper was generally judged to be balanced (Gunther and Schmitt 2004, 64).
The authors argue that this reflects the importance of experts' "imagination
of audience" as a critical factor in their understanding of texts and communications. In this sense, experts are reacting against the media based on
their understanding of the competency and vulnerability of the general
public: "Partisans may believe that information in a mass medium will reach
a large audience of neutral, and perhaps more vulnerable, readers – readers
who could be convinced by unbalanced or misleading information to support the 'wrong' side" (Gunther and Schmitt 2004, 60). In short, Gunther
and Schmitt's research suggests that negative views of the media relate more
directly to experts' views of the general public than to the behaviours of
media institutions themselves.

Experts' negative views of the media are further elaborated by the commentary that we received, which often contain "explanations" of the roots
of media bias. These explanations highlight problems with the media as an
institution, and are particularly critical of the values that guide media actions
and interpretations of expert claims. For instance, both supporters and opponents of aquaculture repeatedly make reference to the media's capitalist
base, albeit in different terms. Respondents who posit that the media have
an anti-aquaculture bias are very plain in their contention that the media
misrepresent the issue because it is in their financial interests to emphasize
the claims made by critics of aquaculture:

> Media people want quick sound bites, not rambling dissertations on uncer-
> tainty. Media people are not unbiased. The media look for and foster con-
> frontation. Controversy is good for media business. (Respondent 146:
> Government Affiliation; Supportive Stance)

The media looks for controversy and as such has to present the "angle" of a story which can be manipulated by people who have vested interests in getting a particular point of view expressed. Environmental groups have a much stronger knowledge of what stories will get the attention of the general media and hence can get their views across with little or no scientific merit. (Respondent 228: Government Affiliation; Supportive Stance)

The media are thus often viewed as institutions that choose to interpret or represent science according to non-scientific values. Their main priority is seen to be profitability before open communication and objectivity, and they are consequently accused of seeking to exaggerate scientific uncertainties in order to create conflict in the public domain.

Experts who see the media as having a pro-aquaculture bias advance notably similar explanations. In these cases, however, the media are not perceived as independent institutions whose interests are tied to readership or viewership. Instead, the critique of the media's capitalist base is grounded in perceived financial and/or ideological ties to corporations that have an interest (material or ideological) in the expansion of aquaculture. Again, the values of objectivity and open communication clash with the values of capital.

The means of communication are so constrained by the interests of large media corporations. These are rarely critical [of the industry] and when they are, large corporate interests have been shown to be quick to silence those voices. (Respondent 272: University Affiliation; Moderate Stance)

[As scientists,] we cannot just rely on Izzy's media machine to interpret and regurgitate "information" [a reference to the late owner of CanWest Global Communications, one of Canada's largest print and television media conglomerates]. (Respondent 300: Environmentalist Group Affiliation; Oppositional Stance)

Here, the media are accused of omitting (rather than misrepresenting) key claims in order to toe corporate lines. They are accused of masking a controversy or denying scientific claims that that are critical of aquaculture. Thus, both critiques seek to explain media bias in terms of the impingement of non-scientific motivations on scientific claims. Scientists on all sides of the debate contend that the media deny their voices, either by omitting them or by overlaying them with the claims of competitors.

Experts' Views of the General Public

The third dimension of experts' views of public participation in aquaculture involves their perspectives on the general public. When we talk about stakeholders and the media, we are referring to specific actors and groups. The general public is much less precise, but despite this it is a potent force in the aquaculture debate. In Chapter 3, we saw that both pro- and anti-aquaculture groups use public opinion polling in efforts to gain broader legitimacy for their positions and claims. Existing research shows, however, that experts are often very skeptical of the abilities of lay members of the general public to understand complex scientific and environmental issues (Wynne 1996, 2001). This view is sometimes termed a "deficit understanding of the public," because adherents assume that members of the general public often react emotionally and irrationally to new problems due to lack of understanding (Sturgis and Allum 2004). The deficit model is exemplified in the extensive work on perceptions of science and risk by Slovic (2000, 184-85). Specifically, Slovic suggests that public opinion on risk issues suffers from four limitations: (1) "people's perceptions of risk are often inaccurate," (2) "risk information may frighten and frustrate the public," (3) "strongly held [public] beliefs are hard to modify," and (3) "naïve [public] views are easily manipulated by presentation format." According to deficit model scholars, these limitations threaten the legitimacy of science because they affect the meaning of science communication. In other words, public opinion is seen as a force capable of beating down good science if not handled with care.

Because of this contradiction, where experts are both covetous and skeptical of public opinion, we were unsure whether to expect positive or negative evaluations of the relationship between scientific knowledge and public opinion. Indeed, the data presented below are at first glance contradictory, and are certainly more complex than the firm opinions on stakeholders and the media discussed thus far.

The survey contained several closed-ended questions that directly addressed views on the relationship between scientific knowledge and public opinion. The pattern of responses to the first question, "Most people's opinions on aquaculture are informed by good scientific information," was almost identical to those discussed previously. The correlation between responses to this question (again scored −2 to +2) and the Stance Scale (scored −14 to +14) is a modest but statistically significant .20, meaning that as respondents demonstrate more support for aquaculture, their evaluation of the public's reliance on "good scientific information" in formulating opinions is generally poorer.

Once again, however, the correlation masks the substantive nature of our findings. Table 4.15 shows the mean scores achieved in response to the question, first according to stance and then according to institutional affiliation. In both cases, as with assessments of the media's handling of scientific-knowledge claims, the variance in responses reflects only different degrees of *very poor* opinions. Whereas a score of 0 represents neutrality, all categories are again firmly in the negative.

Table 4.15

Views on the grounding of public opinion on aquaculture in "good science": mean scores of survey respondents by stance on aquaculture and institutional affiliation

	"Most people's opinions on aquaculture are informed by good scientific information"
Stance on aquaculture	
Opponent	*−1.18*
Moderate	*−0.94*
Supporter	*−1.38*
Institutional affiliation	
University or college	*−1.13*
Government	*−1.13*
Industry	*−1.52*
Environmentalist group	*−0.91*

Notes:

1 For this statement, survey responses ranged from −2 (strongly disagree) to +2 (strongly agree).

2 Differences between the categories based on stance on aquaculture and on institutional affiliation are significant at $p < .05$ (*F*-test).

Some respondents frame this problem as inherently rooted in the cognitive limitations of the lay public:

> The public understands very little about science or technology so they cannot fully understand risk/benefit [analysis]. They basically cannot speak the language, so things have to be dumbed down so much that meaning is lost. (Respondent 255: Government Affiliation; Supportive Stance)

> [Scientists] are closer to the facts ... We [and the public] don't speak the same language. (Respondent 204: Environmentalist Group Affiliation; Oppositional Stance)

Anecdotally, these comments are in marked accordance with Slovic's limitations argument, discussed above. It is equally significant, however, that many scientists go beyond a conceptualization of the public as simply different and incapable of understanding or grasping the principles of science. For them, the public is viewed as a powerful but misguided actor. Thus, some respondents frame their frustrations according to what they view as poor decision making and misplaced priorities:

> The public views things as black and white. In some cases, they refuse to have any risks placed upon them from outside no matter what the tradeoffs

may be. On the other hand, the public in general are constantly choosing risky behaviors for themselves with little compunction. (Respondent 259: University Affiliation; Moderate Stance)

Joe and Josephine Q. Public will only hear that having [an aquaculture] farm in their bay will have the same net result as having x-thousand toilets flushed into the cove per day (or whatever), yet they want cheap fish. Go figure ... They hear what affects them directly (for example the "sanitary health" of their locale or the resale price of their cottage) and not those effects more loosely associated with their daily lives, such as employment or the price of fish. (Respondent 104: Government Affiliation; Moderate Stance)

The public still sees the Neolithic revolution as a positive occurrence for humanity – which was not the case. Aquaculture is one of the latest mani-festations of our desire to control what should not be controlled. (Respondent 138: University Affiliation; Oppositional Stance)

Importantly, this type of commentary is not about science but about values. Although scientists also attack the media with respect to values, such critiques are based on ideas about the media as an interested institution that superimposes outside considerations and priorities onto scientific claims and communications. In contrast, when speaking about the general public, scientists decrease their emphasis on expert claims and instead criticize the values that lead the public to adopt certain positions or behaviours:

There is sufficient information available so most people can make informed decisions. [But] they may not do this, and [instead] choose to rely on unreli-able sources. (Respondent 114: University Affiliation; Moderate Stance)

In short, although experts are highly critical of the lay public's reception, consumption, and interpretation of expert knowledge, our respondents do not frame this critique in terms of a public rejection or willful misinterpreta-tion of science. The media are guilty of this, but the general public is seen as either simply incapable of grasping the language of science or as being actively misled by interested parties, including the media. As a result, the public is thought to make poor decisions and/or cognitive choices because of misapplied values and misplaced priorities. Assumptions about purposeful misrepresentation, which are absolutely central to the critiques levelled against the media by experts on all sides of the debate, are absent here.

Thus far, we have uncovered three very different attitudes towards public involvement in aquaculture: (1) stakeholder groups (local and Aboriginal communities) are perceived as legitimate if limited contributors to aquacul-ture science; (2) the media are excoriated by all sides for alleged manipulation

Table 4.16

Views on public trust of "the aquaculture science produced in my home organization": mean scores of survey respondents by stance on aquaculture and institutional affiliation

	"The public trusts the aquaculture science produced in my home organization"
Stance on aquaculture	
Opponent	+0.41
Moderate	+0.31
Supporter	+0.34
Institutional affiliation	
University or college	+0.43
Government	+0.22
Industry	+0.33
Environmentalist group	+0.80

Note: For this statement, survey responses ranged from –2 (strongly disagree) to +2 (strongly agree).

of science communication; and (3) the general public is strongly criticized, but only on the basis of poor decision making and misplaced values. The results from one final question about the general public suggest some common themes and linkages across these findings.

The question involves the statement "The public trusts the aquaculture science produced in my home organization." As with the other statements, responses are scored from –2 (strongly disagree) to +2 (strongly agree). Importantly, the pattern of responses to this question breaks with those found with regard to other closed-ended questions discussed thus far. First, there is almost no correlation between scores achieved on the Stance Scale and agreement or disagreement with this statement (0.03). Second, as demonstrated in Table 4.16, the means achieved across all categories are *positive*. Thus, despite consistent and strongly negative views of the general public's abilities and willingness to base their opinions on science, experts are equally consistent in their assertion that the public trusts the "aquaculture science produced in my home organization." We contend that this question exposes the limits of experts' critique of lay knowledge and actions because it contains a reference to the *relationship* between the respondent and lay persons. This particular question shifts the frame of reference from how public entities receive and use scientific claims to how they perceive and evaluate claims makers.

This exposes a key distinction in experts' understanding of the public. We are now in a position to directly contrast the overwhelmingly negative evaluations of the media and of public opinion with the strongly positive views on stakeholder participation and perceived public trust in experts' own organizations. We argue that experts on all sides of the controversy make poor evaluations of the ways in which lay entities formulate or make claims to knowledge. They condemn what they perceive as the media's intentional misuse of expert knowledge to claim authority on aquaculture, and they criticize the incompetence that leads the general public to simultaneously interpret and ignore scientific claims when reaching opinions on the issue. *These critiques are about the lay handling of scientific knowledge.*

In contrast, experts evaluate positively lay actors and activities when these compliment or complement traditional science. Experts' positive views on stakeholder participation are shown in the open-ended commentary to have limits. Experts in aquaculture are very open to the participation of the "community" in regulation and/or policy, but in well-defined roles that support scientific practices and processes instead of challenging them. In the words of one respondent, stakeholder inputs are factors "that need to be considered next." The positive responses to the statement "The public trusts the aquaculture science produced in my home organization" can be similarly interpreted. The statement asks experts to imagine a view of themselves through the eyes of others. The language of "trust" as opposed to "use" or "interpretation" places the onus *back onto the claims and the claims makers themselves.* In this context, it is not surprising that experts on all sides of the controversy are reluctant to acknowledge any disjuncture between public trust and the work of one's home organization. To do so would call into question the legitimacy of one's own claims.

When taken together, our findings present something of a paradox: in the struggle over aquaculture knowledge, experts both deny public opinion and claim it. The denial is based on the effect that lay reception and use of expert knowledge has on the meaning of that knowledge. The claiming of public opinion stems from both a genuine desire to engage stakeholders in science and a need for the legitimacy and validation granted to expert claims through this process. In the first instance, control over claims and knowledge is lost; in the second, it is gained and reinforced. Both trends – the positive evaluations of some forms of public participation and the negative views of others – are integral to the complex portrait of expert understandings of the public that emerges from our findings.

Conclusion

We began this chapter by stating that if the aquaculture controversy is a knowledge battlefield – where pro- and anti-aquaculture activists publicly struggle over whose claims and interpretations are most valid – then it is important to look carefully at the claims makers. Are aquaculture scientists and other experts indeed "knowledge warriors" who are deeply divided and actively engaged in the fight over the future of the industry? Or is there more consensus within the expert community than is evident in the exchanges among activist groups?

Overall, our findings answer both questions yes and no. The survey shows that the experts are deeply divided on the risks and benefits of aquaculture. The Stance Scale, which measures fourteen dimensions of experts' assessments of aquaculture, turned out to be highly valid both statistically and substantively. Experts are found across the entire range of the scale, indicating an extraordinary variety of professional opinions about aquaculture. Within the scale, however, we find that some areas of disagreement are more extreme than others. For instance, experts showed more agreement on the economic benefits of aquaculture than we expected, whereas there were deep disagreements on the environmental and human health dimensions.

The most suggestive finding involves the patterning of different opinions on aquaculture. As we saw, much of the variance on the Stance Scale can be explained by knowing an expert's institutional affiliation, professional network, and personal values. The strength of these associations led us to speculate that experts' professional and personal situations contribute to particular lenses through which they view aquaculture. That is, aquaculture looks different to experts embedded in different networks and holding different worldviews. There is undoubtedly an element of self-selection at work here, where people seek out likeminded others in their careers and networks. Nevertheless, this finding suggests the possibility that multiple "expert communities" on aquaculture have developed that have very different perspectives on its risks and benefits.

Our findings regarding experts' views of one another are similarly complex. Aquaculture experts on all sides of the debate are frustrated, and this frustration is expressed in two ways. First, it is expressed in attacks on one another. As we saw, mistrust is a serious problem. Most tellingly, with the exception of high levels of trust in academia, experts are deeply mistrustful of those who disagree with them. This suggests that battle lines have been drawn. The second way in which frustration is expressed is through externalization of the problem of mistrust. As we saw, many experts blame outsiders and outside influences (particularly politics) for introducing bias. This is an important argument, because it involves a closing of ranks around the expert community as a whole.

Finally, our findings regarding experts' views of the public reinforce this defence-of-science position. Experts on all sides of the controversy view public participation in the controversy in essentially the same, if somewhat contradictory, way. That is, experts are highly critical of lay interpretation and use of expert knowledge, while remaining open to the limited participation of the public in fundamental expert activities such as risk assessment and policy. We argued that this is consistent with a defence-of-science position. Experts are frustrated by the interpretation of aquaculture by outside actors (in this case, the media and the general public), but are willing to let outsiders in – as long as they do not call into question the ultimate authority of expertise.

Altogether, the survey leads us to conclude that aquaculture experts may be knowledge warriors in some respects but not in others. Divisions are clearest on questions regarding the risks and benefits of aquaculture, and in the mistrust that many experts feel towards those who disagree with them in that regard. On many other issues, experts differ only in the intensity of their agreements or disagreements (particularly in terms of personal values and views of the general public). Moreover, the open-ended commentary tells us that many knowledge warriors on all sides are more defensive than offensive, more concerned with protecting the integrity of science from the turbulence caused by the controversy than they are with attacking adversaries.

5
Media and the Knowledge Battlefield
with Mary Liston

Media organizations are important players in the aquaculture controversy in Canada. As we saw in Chapters 3 and 4, supporters and opponents of aquaculture are engaged in protracted struggles over knowledge production and communication that often involve media strategies. This is no accident, as media organizations are influential agenda setters in public debates over environmental issues. In the words of Mazur and Lee (1993, 682), "while the news media do not tell us what to think, they are successful in telling people what to think about." That is, while media coverage of aquaculture issues does not dictate public opinion, it does strongly affect how people think about the industry and controversy – by presenting certain arguments, images, and packages of knowledge and meaning for the general public to consume and evaluate (Gamson and Modigliani 1989).

At the same time, however, the media are unpredictable actors in the aquaculture controversy. Media organizations have their own interests to advance. The news media in Canada are overwhelmingly a for-profit endeavour, which means that news production is undertaken with paying audiences and advertisers in mind (Bell 1991). There are also internal interests and pressures at work. The vast majority of journalists are not scientists. Even in cases where journalists have a strong background covering scientific and environmental issues, news-cycle and deadline pressures make it difficult to develop deep understanding of new data and expert claims (Hannigan 2006, 86). Sociological and anthropological studies of newsrooms tell us that journalists often use fallback techniques in order to cope, including relying on third-party information (such as press releases from interested parties) to fill gaps in stories and meet deadlines (Stocking 1999).

In this chapter, we look carefully at trends in news media coverage of the aquaculture controversy in Canada, and examine their relevance to the knowledge battlefield over aquaculture facts and narratives. Specifically, we consider the findings of an analysis of over 1,500 articles about aquaculture that appeared from 1987 to 2006 in seven Canadian daily newspapers.

An analysis on this scale enables us to identify key trends in news coverage, including who is speaking in the media, what is being said, and what is being left out of different narratives about aquaculture. As mentioned in Chapter 4, a quick examination of print media coverage shows relative balance, with most news items adopting a point/counterpoint approach that presents pro- and anti-aquaculture claims and voices together. The more detailed analysis presented in this chapter finds important trends in media coverage, however. Although we do not find evidence of bias in media treatment of aquaculture, we do conclude that media coverage is highly selective. That is, the media tend to focus on particular voices and discursive or thematic "packages" that tell particular types of stories about the industry and the controversy. As we shall see, some tendencies appear to favour the industry and its supporters, while others result in very negative coverage.

The Media and Scientific Controversies

The media play a complicated role in public controversies, particularly in conflicts over scientific knowledge and environmental risk (Friedman et al. 1999). Media have multiple roles in reporting on science and the environment. First and foremost, the media have an investigative role. While most media stories involve more "reporting" than active "investigation," the idea that media serve as watchdogs in the public interest is central to journalists' professional ideology (Sparrow 1992, 2). Investigative reporting is particularly important in environmental journalism. In an analysis of Canadian environmental reporting, Einsiedel and Coughlan (1993, 138) found that "environmental beat writers were more likely to write longer pieces, more analytical pieces, more likely to generate a story on their own rather than depend on an external source to initiate the story, and more likely to challenge conventional institutional wisdom."

The second role of media is communicative, which is again particularly important in scientific and environmental journalism, given that many issues and debates are new and unfamiliar to the public. One of the biggest challenges facing reporters on scientific and environmental issues is to accurately convey complicated expert claims and debates in non-technical language (Stocking 1999). This is both difficult and critically important, as the news media are most often the primary source of information for concerned members of the public (Kasperson et al. 1988). Even in the age of Internet citizen-journalism, "it is unlikely that many of the lay public would have become aware of [new issues such as] mad cow disease or the purported dangers of genetically modified foods if it were not for [mainstream] media reportage" (Hannigan 2006, 79).

The third role of the media is narrative. In a world with a near-infinite number of stories that could be told, journalists have to make the case about "why you should care." Journalists use several techniques to create narrative in scientific and environmental reporting. According to Hannigan (2006, 84), narrative is most easily created when dealing with events. This is why "catastrophes are the bread and butter of environmental news coverage, as they frequently involve injury and loss of life ... and sometimes acts of tremendous courage or self-sacrifice" (see also Allan et al. 2001). This event focus is also evident in science reporting. Stocking (1999, 27) points out that journalists often downplay the long, slow, and collaborative processes of scientific discoveries, instead framing these as "breakthroughs" or "triumphant quests" that are presented as shaking up the status quo. Other narrative devices are more controversial. For instance, environmental stories are often told with an element of urgency and fear, particularly when the issue is new and unfamiliar (Leiss and Nicol 2006). Fearful narratives are means of grabbing attention, but they are also means of dealing with uncertainty (Stocking 1999). When issues are new, experts may not have concrete answers regarding potential consequences or hazards, and journalists often press them for statements of conjecture regarding worst-case scenarios that are then reproduced in the media as potential or even likely outcomes (cf. Hartz and Chappel 1997, 30; Dunwoody 1999). Equally controversial is the device of pitting experts directly against one another to create the illusion of rancorous debate, even when there is broad consensus in the scientific community (Dearing 1995). According to Miller (1992), the media's willingness to give equal time to tobacco industry experts who cast doubt on the link between smoking and lung cancer prolonged this controversy in the public sphere despite strong scientific consensus on the issue. McCright and Dunlap (2003) argue that a similar effect is extending the current controversy over human-caused climate change despite strong evidence of consensus among climatologists (see also Boykoff and Boykoff 2004; Carvalho 2007).

The roles of investigation, communication, and narration can conflict with one another. While science journalists rarely have extensive scientific backgrounds themselves, most take the role of science communication seriously and strive for accuracy (Stocking 1999, 31). At the same time, however, the investigative role of news media often prompts journalists to adopt a critical stance towards claims makers. According to Killingsworth and Palmer (1992), science and environmental journalists therefore often use a skeptical or ironic tone when reporting controversial scientific claims and arguments. Scientists themselves are often offended by this behaviour. In a broad survey of natural scientists in the United States, Hartz and Chappel (1997) found that most deeply distrust the media. Among the most common criticisms levelled by scientists about the media are that journalists lack a basic understanding of scientific methods, are more interested in sensationalism than truth, purposely overstate risks, want instant answers despite uncertainties, and make fundamental errors on points of fact (Hartz and Chappel 1997, 30). The findings from our own survey of aquaculture experts in Canada (discussed in Chapter 4) are also clear in this regard: experts on all sides of the debate have a profound distrust in the ability and motivations of media to accurately represent their claims. As discussed earlier, these poor opinions probably stem in part from experts' "imaginations of audience." That is, experts are keenly aware of the potential influence of the media in public debate, and are therefore prone to see bias even when coverage is relatively balanced (Gunther and Schmitt 2004).

The Study

Our analysis of media coverage of aquaculture in Canada is based on the careful reading and coding of 1,558 print media items appearing in seven major Canadian newspapers between January 1987 and December 2006. Sources for the study include Canada's two main national newspapers (*Globe and Mail* and *National Post*), three Pacific region newspapers (*Vancouver Sun, The Province,* and *Victoria Times-Colonist*), and two Atlantic region newspapers (*Halifax Daily News* and *St. John's Telegram*). These particular newspapers were selected for several reasons. First, we deemed it important to examine media from aquaculture-intensive regions (Atlantic and Pacific coasts) as well as media with national reach. Second, we sought to include both broadsheet and tabloid-format media; the latter generally favour shorter articles, simpler language, and more attention-grabbing editorial strategies (cf. Schaffer 1995; Connell 1998). The *Halifax Daily News* and *The Province* are both tabloid-format newspapers, while the others are broadsheets. Finally, we selected newspapers based on the availability of electronic archives.[1]

The archives of each newspaper were searched using the keywords "aquaculture," "salmon farm," "shellfish farm," and "fish farm." Items that were identified using this method were then individually examined for relevance. In order for an item to be included in our final analysis, aquaculture had to be a major theme in the article. For example, we found that passing references to the keywords were sometimes made in restaurant reviews, reports on stock market performance, and even in obituaries. Items such as these were considered irrelevant to the aquaculture industry or controversy and were excluded. Furthermore, several Atlantic provinces have recently re-named their "Departments of Fisheries" to "Departments of Fisheries and Aquaculture." This sometimes led to spurious selection, and items that did not deal directly with aquaculture issues were screened out of the final analysis. One final caveat: at the time of our research, not all newspapers had electronic archives going back to 1987. The *Globe and Mail* and the *Vancouver Sun* have complete records, *The Province* goes back to 1989, the *Halifax Daily News* to 1991, the *Victoria Times-Colonist* to 1993, and the *St. John's Telegram* to 1997. The *National Post* has complete records, but began publishing only in 1998. We will take these limitations into account in our later analyses of longitudinal trends.

The study enables us to look for trends in media coverage across several dimensions. First, it is a means of examining who is speaking about aquaculture in the print media. As we have seen throughout this book, the aquaculture controversy in Canada is complex and involves a wide range of actors, from the local residents of coastal Canada to organized industry lobbies and professional environmentalist groups. The media study can tell us whose voices are being heard in the public discourse about aquaculture. Second, the study is a means of investigating key themes in media coverage. Each of the 1,558 items has been coded for mentions of different dimensions of the controversy (such as mentions of various risks or harms, benefits, or types of knowledge claims; see Table 5.2). Using statistical analysis, we can identify which types of argument tend to go together in media items, and, alternatively, which are rarely found together. This enables us to look at what Gamson and Modigliani (1989) call "the discursive packages" that are used in media coverage – the sets of ideas and claims that are presented together as narratives. Finally, the study enables us to look at variance in media themes and voices across time and regions. Issues typically rise and fall in the media as part of what Downs (1972) has called "issue attention cycles," and we find clear evidence of this with the aquaculture controversy in Canada. We also expect media coverage to differ across Pacific, Atlantic, and national media, as these speak to different audiences. By considering each of these dimensions, we develop a broad analysis of aquaculture reporting in Canada.

Findings

Our analysis of the 1,558 articles shows several important patterns in media coverage of aquaculture. First, we will discuss fluctuations in the volume of coverage of aquaculture over time. As we shall see, media coverage of aquaculture is sometimes more and sometimes less intense, but since the 1980s aquaculture has never been entirely out of the news. Second, we will look at trends in the themes and arguments raised in media coverage of aquaculture. Specifically, we will look at which themes and arguments are most common and least common in media discourses. Third, we will use statistical analysis to look at interrelationships among themes in order to identify "discursive packages" that are being used in the media. Fourth, we will look at who is being cited in the media, and what is being said by them. At that point, we will also address an inconsistency in our findings: anti-aquaculture themes of risk and harm appear more often in media coverage than do pro-aquaculture themes, at the same time that industry and pro-industry voices are heard more often than any others. Last, we will look at some of the key regional variations in the themes and voices implicated in aquaculture coverage.

Figure 5.1

Number of items on aquaculture appearing in the *Vancouver Sun*, 1987-2006

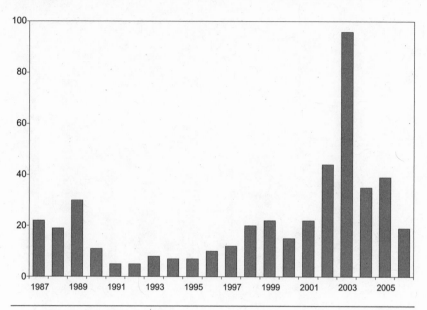

Volume and Type of Coverage

Media coverage of aquaculture has fluctuated since the industry first began in earnest in Canada in the mid-1980s. For example, Figure 5.1 shows the volume of items appearing in the *Vancouver Sun* from 1987 to 2006. As we can see, coverage was relatively high in the late 1980s, with around twenty articles being published each year. This was followed by a period of abeyance in the early 1990s, when between five and ten articles appeared annually. The number of articles picked up again in the late 1990s, followed by an unprecedented upsurge in the early years of the new millennium.

There are several things to note about these trends. First, the periods of high coverage coincide with the emergence of "problems" within or with the industry. That is, coverage is less (but not absent) when the industry is stable or the controversy is quieter. For instance, the high coverage in 1989 was due to the convergence of several negative storylines about aquaculture, including the devastation of several farms from disease outbreaks and unforeseen marine algal blooms, a glut on world salmon markets, and the bankruptcy of several high-profile firms. Similarly, coverage increased in 1999 due to reporting on large escapes of Atlantic salmon in Pacific waters, as well as the crash of wild Pacific salmon fisheries (that year saw the lowest

Figure 5.2

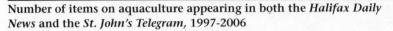

Number of items on aquaculture appearing in both the *Halifax Daily News* and the *St. John's Telegram*, 1997-2006

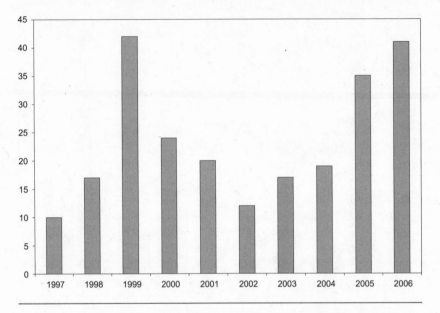

commercial catch of Pacific salmon then on record in Canada). The explosion of coverage in 2003 was due primarily to the publication of claims of sea lice transfer from farmed to wild salmon (see Chapter 3).[2]

A second thing to note about these trends is that although coverage is highest in periods of difficulty for the industry, not all coverage during these times is negative. Many pro-aquaculture items appear during these peak times as well, including opinion pieces, letters to the editor, and articles regarding the economic benefits of aquaculture. For instance, at the height of the sea lice controversy in British Columbia in 2003, positive articles appeared under headlines such as "Blue Revolution Could Net a Profitable Future for BC's Fish Farms" (*Vancouver Sun*, 31 October 2003), and "Looking for Green in the New Blue Revolution" (*The Province*, 14 August 2003). The third significant trend in Figure 5.1 is that aquaculture coverage never quite goes away. In 1991 and 1992, the years of lowest coverage, the articles that did appear were strongly critical, including articles on escapes, disease, and environmentalist protests against the industry.

News media coverage in Atlantic Canada follows the same general pattern of cyclical media attention. Importantly, however, the peaks and valleys in the issue attention cycle occur at different times, which suggests that Pacific and Atlantic media coverage are motivated by different regional issues.

Figure 5.3

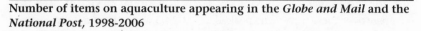

Number of items on aquaculture appearing in the *Globe and Mail* and the *National Post*, 1998-2006

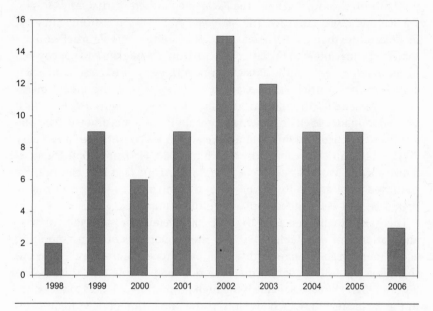

Volumes of coverage in the *Halifax Daily News* and the *St. John's Telegram* are given in Figure 5.2. The graph covers the period from 1997 (the earliest date of electronic access to the *Telegram*) to 2006. In this case, the peak in 1999 was due to a combination of negative and positive events for the aquaculture industry. On the negative side, media coverage focused on disease outbreaks (independent of the Pacific outbreaks), the threat of aquaculture escapes, and several conflicts between shellfish aquaculture and salmon aquaculture operations on the issue of pollution. On the other hand, the media also focused at this time on decisions by provincial governments in the region to fully endorse aquaculture as a means of economic development, which were generally portrayed positively. The peak in coverage in 2005-06 reflects numerous articles on sea lice and other threats to wild Atlantic salmon, as well as generally positive coverage given to aquaculture research being conducted in Atlantic Canada and to government plans to expand the industry. Particularly in Newfoundland and Labrador, considerable attention was being paid to the possibilities for cod aquaculture. As with the Pacific media, periods of intense coverage mean heightened attention to *both* the negative and the positive aspects of the industry.

Fewer articles have appeared in the national newspaper media than in the Pacific and Atlantic media. Figure 5.3 gives the media items appearing in the *Globe and Mail* and the *National Post* from 1998 (the first year of publication for the *National Post*) to 2006. For the national media, we see that coverage over these nine years peaked in 2002-03, when media items focused primarily on the same issues as the Pacific newspapers, particularly escapes, sea lice, and disease transfer. The *National Post* has taken a stronger interest in aquaculture than the *Globe and Mail,* publishing forty-five articles on the subject of aquaculture to the latter's twenty-nine. While the *Post* has covered major developments in the aquaculture controversy, it has also used aquaculture in several articles as an example of the excesses of the environmentalist movement. For instance, in 2003 it ran a series of lengthy articles critical of the dominance of environmentalist groups in the aquaculture debate (the articles were titled "An Enlightening Lesson in Social Justice" [28 February 2003, A14], "An Upstream Battle" [10 June 2003, FP7], and "Soft Totalitarianism of the World's NGOs" [4 July 2003, A16]). These articles notably portrayed local aquaculture workers and companies as victims of highly organized and well-funded anti-aquaculture campaigns.

There are also differences in the types of coverage across Pacific, Atlantic, and national media. Table 5.1 shows the proportions of items appearing across four categories: news articles, editorials (guest or in-house), letters to the editor, and features or information series (long, exposé-type articles). The biggest difference across media is the high proportion of letters to the editor in the Pacific region (which also lowers the proportion of news items compared with Atlantic and national media). As we have discussed several times in this book, the aquaculture controversy is most intense on Canada's Pacific coast. In the Pacific media, one letter to the editor appears for every four news articles, compared with a ratio of 1 to 38 in the Atlantic region and 1 to 55 in the national media. In British Columbia, members of the general public, as well as interest groups, are more willing to write in about aquaculture, and media outlets are more willing to publish these comments.

Interestingly, however, the national media publish proportionally more editorials than the regional media: 1 editorial for every 3.9 news articles,

Table 5.1

Proportion of total media items about aquaculture, by region

	Region		
	Pacific	Atlantic	National
News article (%)	68	84	74
Editorial (%)	12	14	19
Letter to the editor (%)	16	2	1
Feature or information series (%)	4	0.5	5
Total articles	1,212	272	74

compared with ratios of 1 to 5.5 for the Pacific regional newspapers and 1 to 6.2 for the Atlantic regional newspapers. In short, the national media appear to be less interested in daily or weekly news about the aquaculture industry or controversy, but they do not hesitate to address editorial angles of the story.

What Is Being Said?

Journalism involves making choices. Along with their editors, journalists choose what to write about, the theme or angle of the story, the structure of the story, and the sources that they will use (Hansen 1993). These choices are typically guided by professional ethics, prior knowledge, and convenience (Stocking 1999). Although different choices are made by different individuals working for different media organizations, this study enables us to look for broad trends in these choices. That is, we can look for common ways in which the aquaculture story is told in the print media – what is being said or not said in journalists' attempts to explain the issues to their audiences.

To perform this analysis, we coded each of the 1,558 media items according to a standard set of content codes.[3] Table 5.2 presents some key findings from this coding. The frequencies in this table refer to the number of articles that contain mentions of the various codes, rather than to the overall number of mentions. That is, if an article contains five separate mentions of an environmental risk of aquaculture, the article is still counted only once. Table 5.2 therefore enables us to compare the popularity of various themes in media coverage as they appear across different items.

First, we see that media coverage of aquaculture is overwhelmingly preoccupied with finfish aquaculture (which, in Canada, is overwhelmingly salmon aquaculture). Shellfish aquaculture, which is also a major activity in Canada, is mentioned in only 182 articles (12 percent of all articles), compared with 1,210 articles that mention finfish (78 percent of all articles). A further 275 articles do not explicitly indicate which type of aquaculture is being considered, although in most cases we can say that the article is implicitly about salmon aquaculture.[4] The Atlantic media were proportionately more likely to give coverage to shellfish aquaculture issues (mentioned in 27 percent of all articles, compared with only 7 percent of national media articles and 8 percent of articles in the Pacific media).

There are two probable reasons for the media's overall focus on salmon aquaculture. First, as we saw in the Introduction, salmon aquaculture is bigger business than shellfish aquaculture at the national level – the stakes are higher in terms of revenues and capital investments. Second, the intense and ongoing controversies over salmon aquaculture are attractive to the media. We can see some evidence of this when we look at coverage over time. In the periods 1990-95 and 1996-2000, coverage of shellfish aquaculture was stable, at 14 percent of all items. In 2001-06, however, when controversies

Table 5.2

Number of media items containing content codes

Content code	Number
Type of aquaculture	
Mention of finfish aquaculture	1,210
Mention of shellfish aquaculture	182
Type of aquaculture not explicitly indicated	275
Type of knowledge	
Mention of expert knowledge	601
Mention of local or traditional knowledge	40
Themes critical of aquaculture	
Mention of ecological risk or harm	1,015
Mention of economic risk or harm	295
Mention of social or cultural risk or harm	117
Mention of public health risk or harm	153
Mention of "economic growth through environmental reform"	29
Themes supportive of aquaculture	
Mention of ecological benefit of aquaculture	63
Mention of economic benefit of aquaculture	626
Mention of social or cultural benefit of aquaculture	27
Mention of public health benefit of aquaculture	60
Mention of "ecological modernization"	64
Government	
Mention of current state policy	935
Mention of rights or consultation	163

over disease transfer, sea lice, and escapes in finfish aquaculture became particularly intense, the proportion of articles mentioning shellfish aquaculture dropped to 10 percent. Controversies about shellfish aquaculture do exist (particularly over ocean usage and rights), and these are often considered in the media in all regions. Nevertheless, salmon aquaculture is "where the blood is," and this is reflected in volume of coverage.

Table 5.2 also demonstrates the dominance of expert knowledge in media coverage of aquaculture issues. As we argued in Chapters 3 and 4, science is the primary language of the aquaculture debate in Canada. Only forty media items (less than 3 percent of all articles) contain mentions of local or traditional knowledge. A mention of local or traditional knowledge was attributed to any statement, explanation, or claim that is based primarily on personal or group experience. For instance, the following statements were coded as mentions of local or traditional knowledge: "Aquaculture is putting our traditional food sources in jeopardy – the effluent from fish farms is starting to poison our clam beds," and "Residents say they know what's really happening, because it's happening in their own backyards." In Chapter 2, we saw that local and traditional knowledge is a key way of "knowing" about aquaculture and its effects. Coastal people have long relationships with local environments, and local support or opposition to aquaculture is often based on knowledge from such relationships. Despite this, mentions of local knowledge in the print media are rare, and journalists clearly prefer to use science and expertise in order to explain aquaculture to their audiences. Later, we shall see that this knowledge imbalance relates to who is speaking in the aquaculture debate.

The coding of media items also enables us to quantify the appearance of themes in the media. Table 5.2 groups these as "themes critical of aquaculture" and "themes supportive of aquaculture," although it is important to note that the themes are also paired in the following ways:

- *Mention of ecological risk or harm/Mention of ecological benefit*
  Many opponents of aquaculture argue that the industry is doing irreparable harm to environment, wild aquatic species, and ecosystems. A variant of this argument is that aquaculture puts these at risk of future harm. In contrast, supporters of aquaculture often argue that there are ecological benefits from the industry, particularly in taking pressure off endangered wild fish stocks. An example of a mention of ecological risk or harm in the media is the following statement: "Wild salmon migrating to the open ocean are being killed off at an alarming rate by sea lice produced by West Coast fish farms, new research at the University of Alberta suggests. In some instances, the parasites are massacring up to 95 per cent of juvenile fish swimming through inlets where farms are located, says the study" (*Vancouver Sun*, 3 October 2006, B5). In contrast, examples of a mention of ecological benefit are found in the following statements: "Salmon farmers are making a significant contribution to the enhancement of wild salmon stocks, particularly those that are endangered because of overfishing" (*Vancouver Sun*, 23 February 1989, A18), and "Experts at the Ocean Science Centre (OSC) see a number of benefits to farming Atlantic cod [in Newfoundland] ... 'I think that people in Newfoundland would actually welcome a few cod escaping,' said Joe Brown, interim director of the OSC" (*St. John's Telegram*, 3 August 2003, A3).
- *Mention of economic risk or harm/Mention of economic benefit*
  Supporters of aquaculture often assert that the industry is crucial for the development of local, provincial, and national economies. Both the Atlantic and Pacific coastal regions of Canada have been hard-hit by downturns in traditional resource economies, and aquaculture is presented as a way forward. On the other hand, opponents of aquaculture argue that

the industry harms local economies, primarily by further harming traditional industries (such as commercial fishing) and damaging emerging industries (such as ecotourism). An example of a mention of economic risk or harm is the following statement: "[Aquaculture] creates a glut of farmed salmon on the market, and this is making fishermen fish more just to stay afloat ... [And while] fish farmers [recently received] a $10 million bailout package from government ... no compensation has been paid to fishermen" (*The Province*, 3 July 1998, A35). An example of a mention of economic benefit is the following statement: "The CEO of Fishery Products International says Newfoundland has a bright future as a niche-oriented aquaculture producer ... 'Without aquaculture growth, our company would not be able to build our food business. We are very, very dependent on aquaculture for growth'" (*St. John's Telegram*, 11 February 2004, A3).

- *Mention of social or cultural risk or harm/Mention of social or cultural benefit*
  Opponents of aquaculture often argue that the industry is threatening the social and cultural fabric of coastal regions. On the one hand, it threatens the small-operator tradition of commercial fisheries; on the other, it menaces the cultural foundations of coastal First Nations, particularly in British Columbia. In contrast, supporters often argue that aquaculture will enable communities to retain and even enhance their health and vibrancy by providing stable employment in isolated places. An example of a mention of social or cultural risk or harm is the following statement: "Heiltsuk [First Nation] leaders fear the [fish] farms will destroy their way of life, which relies heavily on the wild fishery. 'We all live off the land,' said hereditary Chief Edwin Newman. 'We don't want fish farms to come in here and dump their garbage all over our land'" (*The Province*, 15 January 2003, A10). An example of a mention of social or cultural benefit is the following statement: "Like the recreational and commercial fisheries, the job opportunities aquaculture provides are a bridge to healthy communities, allowing people to invest in their communities and their future" (*Vancouver Sun*, 13 December 2006, A19).

- *Mention of public health risk or harm/Mention of public health benefit*
  Opponents of aquaculture have argued that farmed products contain dangerously high levels of toxins and other contaminants that threaten the health of aquaculture workers and/or consumers. Supporters have countered that consuming aquaculture products reduces the risk of cardiovascular disease and other health problems. An example of a mention of public health risk in the media is the following statement: "Contaminants such as PCBs, dioxins, dieldrin and toxaphene are ... up to 10 times higher in farmed salmon than in wild salmon ... The bottom line for consumers is: Stop eating farmed salmon until salmon farmers clean up their acts" (*Globe and Mail*, 14 January 2004, A19). An example of a mention of public health benefit is the following statement: "Food experts around the world

recognize salmon – farmed or wild – as having a number of health benefits associated with high levels of omega-3 fatty acids, [including] reducing rates of heart disease, stroke, high blood pressure, depression, Alzheimer's disease, and childhood asthma" (*National Post,* 11 February 2001, A17).

- *Mention of "economic growth through environmental reform"/Mention of "ecological modernization"*
  On the first point, some opponents of aquaculture claim that the industry could become more profitable if it radically reformed (for instance, by moving from net-pens to closed-containment systems). An example of this theme in the media is the following statement: "Unlike traditional salmon farming, [these] fish are grown in concrete tanks on land ... The downside of the operation is the huge production cost, [but] the premium didn't worry Alex Campbell Jr., fish buyer for the Thrifty Foods grocery chain. 'They said consumers wouldn't pay for organic produce, too.' The 18-store chain hopes to phase out traditional farmed salmon completely, he said" (*The Province,* 16 June 2002, A6). On the second point, some supporters of aquaculture argue that the industry has greatly cleaned up its act (it has "ecologically modernized") and will continue to do so through incremental rather than radical change. An example of this theme appearing in the media is the following statement: "'The industry has made its mistakes, but it is a growing industry. It doesn't matter what industry it is, you're going to make mistakes going along,' [says Peter Sawchuk]. 'There were a lot of mistakes made 20 to 25 years ago' ... [but], properly managed, he says, the fish farming being conducted today has 'absolutely no impact' on the environment" (*Vancouver Sun,* 2 May 2002, C5).

There are notable differences in the frequency of these themes in media coverage. As demonstrated in Table 5.2, the theme of ecological risk or harm appears most frequently in media items (in 65 percent of all articles). In contrast, mentions of possible ecological benefits of aquaculture appear in only 4 percent of items. Clearly, the media are focusing on environmental risks as a crucial component of aquaculture coverage. Although other risks are also mentioned, these are less common. For instance, economic risks are mentioned in 19 percent of articles, public health risks in 10 percent, and social or cultural risk in 7.5 percent (we will consider the interrelationships between these themes in the next section). The theme of "economic growth through environmental reform" was least common, being mentioned in only 2 percent of articles.[5]

In contrast, the most common pro-aquaculture theme is economic benefit. Mentions of the economic upside or potential of aquaculture appeared in 40 percent of all media items. No other supportive theme receives widespread mention: ecological benefits (4 percent), social or cultural benefits (2 percent), public health benefits (4 percent), ecological modernization (4 percent). This suggests that pro-aquaculture arguments are being articulated in the media almost exclusively in terms of economics. In our view, this ought to be troubling to the industry and its supporters. As seen in Chapter 3, the industry has been subject to repeated environmentalist campaigns highlighting environmental and human health risks, and pro-aquaculture groups and individuals have chosen to respond by issuing strong point-by-point rebuttals. What these media frequencies suggest, however, is that journalists are covering the environmental, health, and cultural terrain with anti-aquaculture themes, and presenting pro-aquaculture positions in the context of economic considerations alone (we will return to this point shortly).

Discursive Packages About Aquaculture

The frequency with which themes appear tells us a lot about media coverage of aquaculture. Equally important, however, are the ways in which these themes interact (or fail to). Journalists tell stories by bringing multiple themes and voices together into what Gamson and Modigliani (1989) call "discursive packages." These packages are important to the construction of different angles or storylines about an issue. For instance, it is common in environmental journalism for mentions of environmental risk and economic benefits to be packaged in the same article in order to create a narrative of "environment versus economy" or "the dark side of industrial development" (King and McCarthy 2005, 139). Discursive packages often combine explanation and interpretation (Gamson and Modigliani 1989, 4), which means that the coincidence of themes can reveal a lot about how stories are being told in the media.

Different journalists and media outlets will choose to highlight different aspects of the controversy in their coverage. Using the content codes discussed above, however, we can look for broad tendencies across media coverage of aquaculture. We begin by cross-tabulating each thematic content code identified in Table 5.2 with all others. We then apply a statistical test called a two-tailed *phi* test, which enables us to uncover any unusual tendencies for a given theme to coincide with other themes in the same articles, as well as any unusual tendency for the theme to *rarely* appear with certain others.

Discursive Packages About Risk or Harm

We first consider discursive packages about risk or harm. As we saw earlier, media coverage of aquaculture often contains mentions of the ecological, economic, health, and social/cultural risks or harms of aquaculture. The frequencies (as presented in Table 5.2) do not tell us, however, how these themes relate to one another. That is, we don't know whether there is a tendency for discussions of these different types of risk to be packaged together, whether they are typically considered separately, or whether they have an aversion to other key themes. These are important questions because they reflect the type of story that is told. As we saw in Chapter 3, the aquaculture industry in Canada has suffered a series of bad news events across several fronts, including escapes of farmed fish (an ecological risk or harm), the discovery of trace organic pollutants in farmed fish (a public health risk or harm), and accusations that aquaculture is making it more difficult for commercial fishers to earn a living (an economic risk or harm). A worst-case scenario for the industry would be if these complaints were commonly presented as a combined package in the media. For instance, if an article on accusations of PCB contamination of aquaculture products is bad news for the industry, it is worse news if the article also considers ecological, economic, and social or cultural risks.

Table 5.3 considers the tendency of the individual risk-or-harm themes to appear in the same article alongside each of the other themes (see Table 5.2 for the complete list of themes). Only themes with extremely strong coincidences and aversions are included in the table (those with a statistical significance on the *phi* test of 99 percent or greater). Looking at the first row after the headings, for instance, we see that mentions of ecological risk or harm have an unusual tendency to coincide with mentions of public health risk and social or cultural risk, as well as an unusual *aversion* to mentions of the benefits of aquaculture, specifically the possible economic, social/cultural,

Table 5.3

Appearance of risk or harm themes with other themes in media items

Theme	Likely to appear with	Not likely to appear with
Ecological risk or harm	Public health risk or harm Social or cultural risk or harm Current state policy Expert knowledge claim	Economic benefit Social or cultural benefit Public health benefit
Economic risk or harm	Social or cultural risk or harm Economic benefit	Public health benefit Public health risk or harm Expert knowledge claim
Public health risk or harm	Ecological risk or harm Public health benefit Current state policy Expert knowledge claim	Economic benefit Economic risk or harm
Social or cultural risk or harm	Ecological risk or harm Economic risk or harm Rights Traditional knowledge claim	–

or public health benefits of the industry or its products. Indeed, the statistical analysis demonstrates that an article containing a mention of ecological risk is 2 times *more likely* than other articles to contain a reference to public health risk, and 51 times more likely to contain a mention of social or cultural risk. Conversely, the statistical analysis demonstrates that articles mentioning ecological risk are 1.8 times *less likely* than other articles to contain a reference to the economic benefits of aquaculture, 1.9 times less likely to mention public health benefits, and 3 times less likely to mention a social or cultural benefit of the industry.

Thus, Table 5.3 tells us that risk issues are indeed being packaged, with journalists frequently merging multiple risks into a single narrative. The different risk dimensions of the controversy are typically raised in one of two ways. First, they are often mentioned by sources who are quoted in the article. As we will discuss in the next section, environmental reporting often makes strong use of sources to articulate issues. This gives sources a particular power to raise and link different dimensions of the controversy, but the authorship of the journalist is still evident in the selection and presentation of quotes. For instance, in a 2006 article in the *Vancouver Sun* entitled "First Nations Visit Norway to Protest Fish Farms" (31 May 2006, D4), two sources are quoted in order to make explicit links across ecological, health, economic, and social/cultural risks:

We have seen the devastation done to wild salmon stocks in the Broughton Archipelago and other regions of the southern British Columbia coast by escapes and sea lice from the salmon farms, and we don't want the same thing to happen in our area.

– Source A

Pan Fish's operations are putting our traditional food sources in jeopardy ... Clam digging is a commercial enterprise and often the only source of income for our villagers. We want Pan Fish to be 100-per-cent responsible for the environmental and human health aspects of their Canadian operations.

– Source B

The second way in which multiple risk dimensions are raised is through listing. This can take the form of a point-by-point walk-through of risk issues associated with aquaculture, or the quasi-rhetorical listing of questions or uncertainties, as in a 2002 article in the *Vancouver Sun* entitled "Salmon Farming Decision Unleashes a Juggernaut" (6 February 2002, A14):

Will the extensive use of antibiotics to protect farmed salmon lead to new strains of disease that neither wild nor farmed fish can tolerate? Does underwater noise intended to frighten predators damage the hearing of marine mammals? How many seals and sea lions should a farm operator be allowed to shoot? Will the wild salmon fishery, already devastated by plummeting prices as a result of competition from farmed fish, become irrelevant? If disaster does strike will the aquaculture industry be forced to bear the cost? To what extent are we willing to see our wild coast domesticated, as Scandinavia's has been?

Of equal importance is the *directionality* of this thematic bundling. For instance, when we take articles containing mentions of ecological risk as our baseline, we see that 12 percent of these also contain mentions of public health risk and 10 percent contain mentions of social or cultural risk. When the baselines are reversed, the effect becomes very strong: 78 percent of articles that mention public health risks also mention ecological risks, and this figure rises to 90 percent among articles that mention social or cultural risk. These imbalances indicate the dominance of the ecological risk theme. With the exception of economic risks, journalists find it nearly impossible to discuss any risk dimension of aquaculture without referring to potential ecological problems.

This is not good news for the industry because it suggests that the media are adopting a narrative about aquaculture that places strong emphasis on ecological harm over other possible themes. Indeed, the last column of Table 5.3, which deals with aversions among the themes, is also important for understanding discursive packages about risk and harm. Here we see that mentions of risk appear infrequently alongside mentions of benefits. At first glance, these aversions make sense, because they represent different angles on aquaculture. According to media theory, however, mentions of risks and benefits should be coinciding (Gamson and Modigliani 1989, 8). Fairness and balance are important journalistic norms, and journalists often adopt a point/counterpoint format when reporting on controversial issues (Einsiedel and Coughlan 1993). Thus, we would expect that mentions of risks and benefits would appear together. Table 5.3 shows that in some cases they do – mentions of economic risk tend to coincide with mentions of economic benefits, and mentions of health risk coincide with mentions of health benefits. We expected to find this trend more broadly, however, particularly with respect to ecological risk, which is the dominant anti-aquaculture theme in the media (Table 5.2). Moreover, as we shall discuss below, journalists often do quote pro- and anti-aquaculture sources together in the same article. Why are pro-aquaculture voices not responding to claims about risk with their own claims about benefits? We shall return to this puzzle in a moment.

A final noteworthy trend in Table 5.3 involves different evocations of knowledge and of state policies. For instance, mentions of ecological and public health risks tend to coincide with mentions of expert knowledge and current government policy. This is consistent with what we saw in Chapter 3 – that opponents and supporters of aquaculture are engaged in "science wars" over the ecological and health impacts of aquaculture. These wars are characterized by exchanges of claim and counterclaim that can extend over many months and years. Many of these debates involve discussions of state policy, with supporters of aquaculture often pointing to government standards and regulations as a means of deflecting criticism, and opponents typically arguing that current regulations fail to protect the environment and/or the public. Interestingly, these trends do not hold for other types of risk. In fact, claims about economic risk or harm tend to be *averse* to expert knowledge claims. This suggests that claims about economic risk, as articulated in the media, tend to be based on intuitive statements rather than facts or figures from expert disciplines such as economics. Finally, we see that mentions of social or cultural risk tend to coincide with local or traditional knowledge claims. This is because social or cultural risk is often raised by Aboriginal voices (see next section), who typically articulate these risks using local or traditional knowledge claims.

In summary, we can see evidence of two distinct discursive packages or narratives about aquaculture risk in the media. The first involves the conjunction of the themes of ecological risk, public health risk, expert knowledge, and state policy. These typically combine to form stinging critiques of the industry and/or the ability of provincial or federal governments to measure, regulate, or mitigate these risks. A clear example can be found in a 2002 article from the *Globe and Mail* entitled "Ill Salmon Treated with Unapproved Drug" (14 June 2002, A6):

Health Canada is allowing an unapproved drug to treat parasite-riddled salmon on Canadian fish farms, despite toxicity studies that raise red flags about the treatment's safety for consumers. The release of the drug emamectin benzoate, or SLICE, used to combat sea-lice infestations among farmed salmon, is symptomatic of how Health Canada's emergency drug-release program for animals is being abused, according to a critic.

"It's a program that was intended for a specific purpose. I think a good purpose. But it's now being fairly badly abused," said Michelle Brill-Edwards, a scientist who made a noisy exit from the agency several years ago over what she said was pressure to approve unsafe drugs ...

Several toxicology studies of SLICE done on rats, dogs and rabbits show that when ingested at high doses the drug can cause side effects such as tremors, spine and brain degeneration and muscle atrophy. Diane Kirkpatrick, director-general of Health Canada's veterinary drugs directorate, did not deny that some studies show adverse side effects of SLICE on animal models. "But let me tell you that every substance known to man if ingested at a certain level will cause adverse effects. The dose makes the poison," she said yesterday.

The second narrative involves the conjunction of social or cultural risk, ecological risk, local or traditional knowledge, and rights themes. This narrative is much less common than the first, but is still powerful. The target is again industry and government, but often in a deeper way (containing implicit or explicit critiques of industrial development and/or colonialism). An example is found in a 2003 article entitled "First Nations Fear Fish Farms Damaging Ancestral Waters" that appeared in the *Victoria Times-Colonist* (2 June 2003, C1):

> Chief Henry Scow reels off place names in the Broughton Archipelago. He doesn't pause to think. Scow, chief of Gilford Island First Nation, knows them all like the back of his hand. This is a demonstration of intimacy, of familiarity with places his ancestors knew and First Nations children will come to know just as well ...
>
> Sea lice are the latest in a list of worries about what farms may be doing to an area First Nations consider their territory. The issue is about the risk from activities they fear is [sic] disrupting the natural cycle of life. For example, they say that a crash in wild pink salmon stocks takes away food that grizzly bears rely on catching in rivers. Concerns by First Nations include waste from farms fouling the sea bed and outbreaks at farms of an infectious virus which has killed millions of farmed salmon. Demonstrations on the water and evictions notices have been served to farms by native people opposed to the industry ...
>
> Roberta Smith, a councillor with the Tsawataineuk First Nation and a member on the Musgamagw Tsawataineuk Tribal Council's board, said native voices are being underestimated and ignored. Elders hold "phenomenal knowledge" about the region, she said. There is no question in Smith's mind about what the role of First Nations is when it comes to the environment. "We are mandated to protect our lands and our water because they are all part of who we are."

Discursive Packages About Benefits

We now consider discursive packages that appear in the media about the benefits of aquaculture. The coincidences of pro-aquaculture themes with others are summarized in Table 5.4. Some of the data presented in this table are the mirror image of those found in Table 5.3. For instance, the aversions between mentions of the benefits of aquaculture (economic, public health, and social/cultural) and mentions of ecological risk or harm represent the same associations discussed above.

Table 5.4

Appearance of benefits themes with other themes in media items

Theme	Likely to appear with	Not likely to appear with
Economic benefit	Ecological benefit Social or cultural benefit	Ecological risk or harm
Ecological benefit	Economic benefit Social or cultural benefit Public health benefit Expert knowledge claim	–
Public health benefit	Ecological benefit Public health risk or harm Current state policy	Ecological risk or harm
Social or cultural benefit	Ecological benefit Economic benefit Rights Traditional knowledge claim	Ecological risk or harm

Table 5.4 also contains significant new data, however. First, we see that trends regarding different types of knowledge are replicated in discussions of the benefits of aquaculture. Expert knowledge dominates talk about ecology and health but not economic matters, while local or traditional knowledge is often evoked in claims about the social or cultural benefits of aquaculture. This suggests that pro-aquaculture narratives closely mirror the risk narratives discussed above *despite the fact that they rarely overlap* in the same articles. For example, talk about health benefits tends to coincide with mentions of expert knowledge and of state agencies and policy. These are the same tendencies as in risk narratives. The following quotation from a 2004 article in the *Globe and Mail* is typical of articles that contain mentions of benefits/expert knowledge/state policy (10 January 2004, A10):

> "Farmed salmon are laced with toxins." ... After reading [headlines such as] that, many Canadians may decide to stop eating the succulent fish. That would be unfortunate. The risk from eating farmed salmon, if risk there is, is extremely low ... Both Health Canada and the Canadian Food Inspection Agency see no need for Canadians to limit the amount of salmon they eat. The US Food and Drug Administration agrees. In fact, it encourages people to eat more salmon ...
>
> Tiny amounts of chemicals such as PCBs are found in many foods, including milk, eggs, pork and beef. Those trace amounts are not considered dangerous. Indeed, whether PCBs cause cancer in humans is still unproven.

Unless and until scientists produce more convincing evidence that eating farmed salmon regularly can be harmful, Canadian diners should feel comfortable eating that unsurpassed dish.

As with the anti-aquaculture quotation regarding the drug SLICE discussed earlier (*Globe and Mail*, 14 June 2002, A6), this excerpt uses expert knowledge to make an argument about the long-term effects of aquaculture. Both quotations also mention state agencies, although the SLICE article casts government experts and agencies in a negative light while this article presents them as authorities. This distinction is common. Although the state plays a significant part in both pro- and anti-aquaculture narratives, its role and competence are portrayed very differently in each.

Second, Table 5.4 tells us that discussions of benefits also tend to be bundled with each other. Articles that mention an economic benefit of aquaculture are twice as likely as other articles to mention an ecological benefit, and 19 times more likely to mention a social or cultural benefit. As we saw with risk narratives, journalists frequently use multiple sources to list different benefit dimensions. A 1999 article from *The Province* entitled "Fishy Business" provides a good example (16 May 1999, A46):

British Columbia's flagging fishing industry could be given a real boost, once a four-year moratorium on salmon farms is lifted ... Vancouver Island has 80 salmon farms supporting 900 direct jobs and another 1,200 jobs in closely related businesses, says Anne McMullin, executive director of the B.C. Salmon Farmers Association. Lifting the moratorium could double that in 10 years, she said.

It's a prospect that gladdens the hearts of those living and working on the Island.

"Communities have been hit by contraction in mining and forestry and we need something to get investment going again," said Chris Kelly, president of Kelly Transport of Campbell River, which takes Vancouver Island farmed salmon to market. Kelly says Campbell River's troubled economy is reflected in its schools: School population declined by 330 students in 1998 and will fall by a further 350 this year.

While this listing of benefits is common, it is also clear that the *economic benefit* theme is just as dominant in pro-aquaculture narratives as the *ecological risk* theme is in risk narratives. For instance, only 6 percent of articles that mention an economic benefit of aquaculture also mention an ecological benefit (such as reducing pressure on wild fish stocks), and only 4 percent also mention a social or cultural benefit (such as community stability). Once again, however, the findings are very different when the baseline is reversed. Fifty-seven percent of articles that mention an ecological benefit of aquaculture also mention an economic benefit, and the figure soars to 93 percent for articles that mention a social or cultural benefit. Journalists rarely discuss the benefits of aquaculture without casting these in an economic light.

The comparative popularity of ecological risk and economic benefit themes suggests that journalists are adopting an economy-versus-environment approach in their coverage of aquaculture. It is important to reiterate, however, that these themes rarely overlap *in the same article*. According to our coding, mentions of ecological risk and economic benefit overlap in only 358 of the 1,558 articles. This figure is low, given that 1,015 articles mention an ecological risk and 626 mention an economic benefit (the *phi* measure indicates that this aversion is highly significant; see Table 5.3). This suggests that journalists are generally choosing to highlight either negative themes (built around the notion of ecological risk) or positive themes (built around the notion of economic benefits). As mentioned earlier, this tendency is somewhat confusing, because journalistic norms suggest that reporters strive for balance in reporting controversial issues. It is equally unexpected because, as we shall discuss momentarily, pro- and anti-aquaculture voices have a strong tendency to appear *together* in the same articles. In order to further tease out the relationships among themes in media coverage of aquaculture, we turn next to the question of voice.

Who Is Speaking?
Journalists rely heavily on sources to both learn about and narrate an issue, and this is particularly true in environmental journalism (Anderson 1993). Few reporters have scientific training, so sources are often crucial in interpreting complex and unfamiliar issues. Furthermore, in environmental journalism the disagreement is often the story. In this respect, sources are ways of "making narrative" for reporters seeking to make their stories interesting for editors and readers.

Sources also have a strong influence on reporters and the content of reporting, however. Because media production usually operates on deadlines, journalists often use direct quotes from sources to both explain and comment on complex environmental issues (as we saw earlier). This means that sources can become what Hall and colleagues (1978) term "primary definers" of environmental issues. That is, sources can become the lead voices of discovery and agenda setting in a controversy – simultaneously offering explanations and recommendations under the guise of factual reporting. Indeed, while scholars of media and environmental controversies often pay close attention to media campaigns by antagonists (particularly press releases and paid advertising), Anderson (1993) argues that "source wars" are equally important, where industries and environmentalist groups increasingly look

Table 5.5

Number of media items containing quotes from various sources

Source	Number of items	% of all articles
Member of industry or pro-industry lobby group	648	42
Member of opposition group	522	34
Federal or provincial government employee	491	32
Federal or provincial elected person	444	29
University- or college-based expert	165	11
Coastal non–First Nations person	160	10
Coastal First Nations person	123	8

to cultivate long-term relationships with journalists to strategically position themselves as trusted sources on contentious issues.

In this section, we look at who is being cited in articles about aquaculture in Canada, and what they are saying. Table 5.5 provides the number and proportion of articles that directly quote different types of sources. As before, these figures do not represent the number of times that these sources are cited, but rather the number of articles that contain at least one citation from the given category.

Table 5.5 demonstrates the dominance of industry, opposition-group, and government voices in aquaculture reporting. Members of the industry or pro-industry lobby groups are cited most frequently, appearing in 42 percent of all articles concerning aquaculture in our sample newspapers. This dominance is underscored when we exclude those articles that contain no quotes or voices at all (a total of 222 items) – typically shorter items that report minor developments or points of fact. When these are excluded, the proportion of articles containing industry voices rises to nearly 50 percent.

The dominance of industry voices in the media prompts us to look carefully at what is being said in articles that quote them. To do so, we use the same cross-tabulation method and *phi* analysis discussed above (comparing industry voices with all thematic content codes, and looking for unusual tendencies for voices to coincide with a given theme and/or an unusual tendency towards mutual aversion). The findings are summarized in Table 5.6. Using this analysis, we discover that industry voices appear unusually frequently with mentions of the economic benefits of aquaculture and of ecological modernization. An article that contains an industry voice is 1.9 times more likely to contain a mention of economic benefits than an article without an industry quote, and 2.1 times more likely to carry the ecological modernization theme. Not surprisingly, industry voices are talking about the economic upside of aquaculture and the steps taken by industry to improve environmental performance.

What is surprising, at least initially, is that industry voices also appear unusually frequently with mentions of the ecological risks or harms of aquaculture, as well as with mentions of the economic risks posed by aquaculture to other industries. Why are industry voices speaking about the risks of aquaculture? The truth is, they are not. We will show later that industry representatives frequently find themselves *drawn* into narratives about the negative effects of aquaculture. This is important, because it suggests that the dominance of industry voices in the media may be illusory. As mentioned in Chapter 4, our survey of experts found that aquaculture supporters are the *most critical of the media*. If industry voices are heard most often in the media (in the form of *direct quotations*), why do aquaculture supporters have such negatives views? We will return to this question in a discussion of the interactions among voices, arguing that speaking the most does not necessarily mean dominating the conversation.

The second most prominent voice in media coverage is that of opposition groups, including environmentalist organizations and other groups that have taken an explicit stand against aquaculture.[6] Members of these groups are cited in over one-third of all media items in our sample. When no-citation articles are removed from the sample, members of environmentalist or anti-aquaculture groups are cited in nearly 40 percent of articles that contain any source quotes at all. On the one hand, the high profile of these voices is expected. As we saw in Chapter 3, opposition groups have adopted an aggressive communications strategy that involves the release of scientific studies coupled with media guides and press conferences. On the other hand, their ubiquity is extraordinary. Although the anti-aquaculture movement involves large organizations, such as Greenpeace and the World Wildlife Fund, most activism in Canada is carried out by a relatively small network. Journalists often have many motives for quoting activists (including the creation of conflict or narrative), but the success of this network in establishing its members as credible voices regarding aquaculture policy and practices is notable.

Table 5.6 shows that articles containing environmentalist or anti-aquaculture voices also demonstrate unusual thematic tendencies. Using the same statistical test applied to media items containing industry voices, we see that articles quoting members of environmentalist or anti-aquaculture groups are *much more likely* than other articles to contain references to expert knowledge, ecological risk, public health risk, and government policy, and are *much less likely* than other articles to contain references to the economic benefits of aquaculture. The unusually high coincidence of environmentalist voices and mentions of ecological and health risk are expected. The unusual frequency of mentions of expert knowledge is also significant, as it reinforces

Table 5.6

Intersections of voices and themes in articles about aquaculture

Voice	Likely to appear with	Not likely to appear with
Industry or pro-industry lobby	Economic benefit Ecological modernization Ecological risk or harm Economic risk or harm	–
Opposition group	Expert knowledge claim Ecological risk or harm Public health risk or harm Current state policy	Economic benefit
Federal or provincial government employee	Expert knowledge claim Ecological benefit Ecological risk or harm Current state policy	–
Federal or provincial elected person	Economic benefit Current state policy	Expert knowledge claim Public health risk or harm Public health benefit
University- or college-based person	Expert knowledge claim Ecological risk or harm Ecological benefit Public health risk or harm Public health benefit	Current state policy
Coastal non–First Nations person	Traditional knowledge claim Economic risk or harm Social or cultural benefit	Expert knowledge claim
Coastal First Nations person	Traditional knowledge claim Rights Ecological risk or harm Social or cultural risk or harm Social or cultural benefit Current state policy	–

our earlier argument that environmentalist and anti-aquaculture groups are choosing the language of science and expertise to articulate their opposition to the industry. An alternative explanation, which is difficult to verify but may play a small role in this trend, is that the news media are focusing on what activists are saying about scientific issues and ignoring what they are saying about other issues. Other narratives are possible, as we have seen in this book, particularly local narratives about rights, livelihood, and continuity. Whatever the reason, it is clear that opponents of aquaculture are key drivers of "science talk" in the media.

The third and fourth most cited groups in the media are government employees and elected officials (particularly ministers and opposition-party politicians). High-profile environmental controversies often implicate politicians (cf. Trumbo 1996). On the one hand, politicians are frequently drawn into environmental controversies by journalists, activists, and political adversaries, who ask, "What are you going to do about this?" On the other hand, politicians often engage environmental controversies purposely and directly in attempts to frame or spin the issue in a certain way. For instance, the aquaculture industry has enjoyed the support of every provincial and federal government in Canada since its inception. Even the New Democratic Party (NDP) government in British Columbia (1991-2001), which clearly had reservations about aquaculture, nonetheless oversaw its substantial expansion. Thus, many of the political voices appearing in the media are speaking in defence of aquaculture. Using the *phi* measure, we find that articles containing political voices are *much more likely* than other items to contain mentions of the economic benefits of aquaculture, as well as mentions of current government policies to regulate aquaculture. Conversely, political voices are *much less likely* to coincide with mentions of expert knowledge. This is consistent with research by Trumbo (1996) into media voices and environmental controversies: politicians talk much less about science and technical considerations than they do about values, choices, and the political big picture – which in this case appears to be the economic upside of the industry.

Finally, the weakest voices are those of coastal stakeholders. This category consists of residents of coastal communities who are *not* identifiably linked to industry (in a managerial or lobby position), opposition groups, or government organizations. Included in this category are local politicians and other leaders, businesspeople, fishers, aquaculture workers, property owners, and average citizens. Only 8 percent and 10 percent of articles, respectively, contained a quote from First Nations and non–First Nations coastal stakeholders.

The paucity of coastal voices in aquaculture reporting is a consequence of the choices made by journalists and editors. As we have seen, the aquaculture controversy takes place at multiple levels. In Chapter 2, we argued that local understandings of aquaculture (both positive and negative) are grounded in intimate questions of livelihood and rights. In Chapters 3 and 4, we discussed the "organized" conflict between industry supporters and opponents – the strategic struggle, dominated by expert knowledge claims, to gain legitimacy and discredit adversaries on the knowledge battlefield. The privileging of organizational voices and the relative lack of local voices clearly shows that journalists are focusing on the latter struggle. Local voices are rarely found in standard news articles, but appear in 40 percent of longer feature or information series articles. Given that aquaculture is predominantly conducted on the rural coast, it is notable that local voices enter the media discussion of aquaculture only when journalists decide to dig deeper into the issue.

Table 5.6 shows that articles that do feature local voices differ strongly in thematic content from those that do not. Again using the *phi* test, we see that the former are more likely to contain mentions of local or traditional knowledge, the economic harms or risks of aquaculture, and the social or cultural benefits of aquaculture. The coincidence of these voices and local or traditional knowledge is an expected but significant finding. Although local and traditional knowledge do not appear in the media very often, they do appear when local voices are being heard. The mentions of economic risk or harm parallel the fears, articulated in Chapter 2, that aquaculture may damage existing businesses and livelihoods. Similarly, the unusually frequent mention of social or cultural benefits reflects the hope among some local stakeholders that aquaculture can lend some stability and continuity to rural livelihoods.

Media items containing local First Nations voices also differ significantly from those that do not. The 123 articles that directly quote Aboriginal people (see Table 5.5) are much more likely than other items to contain mentions of traditional knowledge, problems of consultation or rights, ecological risks or harms, social or cultural risks, and the social or cultural benefits of aquaculture. The themes of ecological risk and rights are particularly strong (80 percent and 60 percent, respectively, of articles containing Aboriginal voices). The other themes reflect the variety of First Nations perspectives on aquaculture discussed in Chapter 2: depending on the speaker, the industry is perceived as a threat to community and culture or as a potential means of preserving them.

The Dominance and Vulnerability of Industry Voices

Thus far, we have seen that some voices speak much more frequently than others in media coverage of aquaculture, and that these different voices are associated with different themes in the controversy. Among the most important findings is that industry voices appear more frequently than any other. From this, we might assume that industry messages are the most dominant in media coverage of aquaculture. As mentioned earlier, however, there is a difference between speaking the most and leading the conversation. In the previous section, we saw that industry voices strongly coincide with both pro- and anti-aquaculture themes (particularly mentions of ecological risk or harm). This prompts us to examine how industry voices are invoked in media coverage relative to other speakers.

To explore the intersection of voices in media items, we again use the method of cross-tabulation using the two-tailed *phi* test. This enables us to identify which voices tend to appear in the same article with unusual frequency and, alternatively, which tend to appear unusually infrequently together. We find that quotes from industry members and representatives tend to appear in articles that also contain the voices of opposition group members and First Nations persons. The frequency of these intersections is quite high, in fact. Members of industry and opposition groups are quoted together in 258 articles in our sample, meaning that half of all articles citing members of opposition groups also contain quotes from industry members. A similar pattern is evident for First Nations quotations, which intersect with industry voices 65 times (again, over half of all articles in our sample that quote First Nations people also cite industry members).

This indicates that journalists are using techniques of point/counterpoint in order to explain aquaculture to audiences and provide balance (Einsiedel and Coughlan 1993). The use of point/counterpoint appears to be highly selective, however. We saw earlier that mentions of risks and benefits rarely appear together. Instead, point/counterpoint is used *only within discussions of risk or harm*. For instance, articles that contain *both* industry and opposition voices are 8.5 times more likely than other articles to contain mentions of environmental harms or risks,[7] and are twice as likely as other articles to contain mentions of the health risks of aquaculture. Importantly, the effects are not reversed. That is, articles containing both industry and opposition voices *do not* tend to mention the environmental benefits, economic benefits, or health benefits of aquaculture more than other articles. This is a significant finding because it indicates that when multiple voices are present, the discussion is overwhelmingly negative towards aquaculture, and that industry voices, therefore, are typically brought into media articles to rebut critics rather than to lead the discussion.

These trends are replicated in cases where industry and First Nations voices are found together in the same article. In these cases, mentions of environmental risk or harm are 5 times more frequent than in other articles, and mentions of social or cultural risk are 13 times more frequent. Again, this effect is not reversed. Although First Nations voices appear in only 8 percent of all media items about aquaculture, they, like opposition groups, appear to be setting the agenda in articles that also include industry voices.

These findings give us reason to rethink the apparent dominance of industry voices in media coverage of aquaculture. Members of industry and industry lobby groups are clearly successful in speaking through the media. If there is indeed a "source war" at play, as Anderson (1993) suggests is often the case in high-profile environmental controversies, the aquaculture industry is well represented. It is equally clear, however, that industry members have failed to parlay this central source role into status as "primary definers" of aquaculture in the media (cf. Hall et al. 1978). Industry voices are generally able to stress positive themes in the media only when they appear alone. When they appear with other voices, they are invoked chiefly as rebutters of claims against the industry – hardly a position of dominance or strength.

These findings also shed some light on journalistic practices. The point/counterpoint technique is a recognized way of ensuring neutrality (Einsiedel and Coughlan 1993), but our findings suggest that reporters are making decisions on when to use this narrative vehicle. When industry representatives are talking about the benefits of aquaculture, they tend to do so unchallenged. Of all the articles that contain mentions of the benefits of aquaculture, only 16 percent also contain opposition voices. The converse is very different: 43 percent of articles that mention risk also contain industry voices. *Thus, journalists tend to stage debates much more frequently on the risks of aquaculture than on its benefits.* Superficially, this would appear to benefit the industry, as its claims are generally presented uncontested, whereas it is given the opportunity to rebut the claims of opponents. We suggest otherwise: because point/counterpoint discussions about aquaculture are nearly always on the topic of risk and harm, it may give the impression that industry supporters cannot win an argument. At the very least, this trend suggests that the numerical dominance of industry voices in media coverage of aquaculture is illusory.

Regional Variations in Aquaculture Themes and Voices

We saw earlier in this chapter that the volume of media coverage of aquaculture has varied across Pacific, Atlantic, and national media. In this final section, we address the variability of content across these media groups. Although there is a lot of consistency across the regions, there are also some key differences, both in themes and in voice.

We begin by considering several trends that are unique to the Pacific media in our sample (*Vancouver Sun, The Province, Victoria Times-Colonist*). As discussed in the Introduction, the aquaculture controversy is most intense on the Pacific coast. This is due to several factors, including the decline of the commercial fishery for Pacific wild salmon, the aquaculture industry's use of an exotic species (Atlantic salmon), British Columbia's politically powerful environmentalist movement, and unresolved treaty issues with First Nations groups. For these reasons, it is not surprising that the Pacific media pay strong attention to issues of ecological risk or harm, which are mentioned in 73 percent of media items from that region. Seventy-three percent of items in the national media also mention ecological risk, but only 29 percent in Atlantic media (for reasons we will consider in a moment). The second noteworthy trend in the Pacific media is the reduced emphasis on the economic benefits of aquaculture, which are mentioned in 36 percent of articles, compared with 56 percent of articles in the national media and 49 percent of articles in the Atlantic media. The third trend is the relatively strong coverage given to Aboriginal voices and issues of rights. Although Table 5.5 shows that coastal First Nations people are the least quoted of any major source group on aquaculture, their voices appear in 9.4 percent of media items appearing in the Pacific region, compared with only 2 percent in the Atlantic region and 5 percent nationally. Similarly, 11 percent of Pacific media items contain references to rights claims, compared with 5 percent of both national and Atlantic media items.

Coverage in the national media (*Globe and Mail, National Post*) is thematically similar to coverage in the Pacific region. As mentioned earlier, the national media have more coverage of the economic benefits of aquaculture and less coverage of the rights theme or Aboriginal voices. The only remaining difference involves the national media's strong reliance on expert knowledge claims, which appear in 52 percent of their articles, compared with 41 percent in the Pacific region.

The Atlantic media (*Halifax Daily News, St. John's Telegram*) vary most from the other two groups. First, as we saw earlier, the Atlantic media are substantially less preoccupied with ecological risks or harms. Whereas this is *the* dominant theme in the national and Pacific media, it is mentioned in only 29 percent of Atlantic media items. This tendency is also evident in the selection of sources. Although environmentalist or other anti-aquaculture voices appear in 38 percent and 39 percent of Pacific and national media items, respectively, they are cited in only 13 percent of items in the Atlantic media.

Does this mean that the Atlantic media generally have a more positive stance towards aquaculture? The answer appears to be yes and no. On the one hand, more articles in the Atlantic media mention the economic benefits of aquaculture than in the Pacific or national media (56 percent of items, compared with 36 percent and 48 percent, respectively). Also, the Atlantic media in our sample rarely mention public health risks (3 percent of items, compared with 11 percent and 16 percent), or social/cultural risks (3 percent, compared with 9 percent and 5 percent). This suggests that coverage is generally more favourable. On the other hand, we find evidence that critiques of aquaculture are different rather than absent. First, we see that the Atlantic media pay very little attention to some of the key debates that preoccupy the Pacific and national media. The Atlantic media rarely mention health or social/cultural risks, but neither do they mention these as benefits (in 1 percent and 0 percent of articles, respectively). Mentions of ecological benefits are similarly rare (in 2 percent of articles, compared with 5 percent in the Pacific region and 7 percent nationally). Second, we find that the Atlantic media pay strong attention to *economic risks or harms,* which are mentioned in 33 percent of articles, compared with 16 percent in the Pacific media and 20 percent nationally.

This suggests that debate in the Atlantic media is less about environment and more about the possible consequences of aquaculture development for other types of coastal industry. To see whether the intersection of these themes is widespread, we cross-tabulate these themes as they appear across all Atlantic media items. Using the *phi* analysis that we discussed earlier, we find that mentions of the economic benefits and the economic risks of aquaculture coincide with unusual frequency. The economic risks-versus-benefits narrative takes several forms in the Atlantic media. Some articles directly highlight conflict between coastal users, with aquaculture playing the role of disruptor. The following is from a 1999 article in the *Halifax Daily News* entitled "Fish Farm Fuels Fear" (10 May 1999, 4):

> A group of Guysborough County residents fears a fish farm proposed for the Tor Bay area will ruin their way of life. Some local fishermen and people who work in the tourism industry want guarantees from the province that the Steelhead trout farming operation Scotia Rainbow Inc. wants to build will not damage the bay. "There's a lot of people who have been making a living for generations who are at risk here," said Charlos Cove innkeeper Jim Colvin ...
>
> Scotia Rainbow's director of marketing said the proposed fish farm would create hundreds of jobs in the area ... The area is desperate for jobs and the proposed fish farm has stretched allegiances thin, said Conway. "It's caused conflicts between families," she said.

Other articles such as "Aquaculture Fever Hits" (*Halifax Daily News*, 24 March 1999, 20), take a cautionary tone, warning about possible future conflicts and economic costs:

> A storm is brewing in St. Margaret's Bay where I spend my summers, but not one generated by a low pressure system. This brouhaha is about aqua-culture. Aquaculture is one of those seductively simple ideas such as equal pay for work of equal value or affirmative action – easy to say, but devilishly difficult to implement or regulate fairly. Everyone loves aquaculture until they study it. Then, they become cautious. Led by Fisheries Minister Keith Colwell, Nova Scotia is passionately promoting aquaculture ...
>
> If world-wide experiences of salmon and trout farms are any guide, this initiative could turn the bay into a seabound Sysco, using taxpayer dollars to subsidize a marginal industry which, in the end, could cost more than anyone ever expected.

Finally, some articles that tend to be more positive about aquaculture stress the notion of balance between traditional economic activities (particularly commercial fishing) and the aquaculture industry. The following is from a 2006 article in the *St. John's Telegram* entitled "Fish or Fowl?" (16 May 2006, C9):

Will aquaculture be the fishery of the future? [Mike] Rose, [former executive director of the Newfoundland Aquaculture Industries Association], answers that question two ways. "On a global scale, aquaculture is the fishery of the future. You just have to do the math in terms of global population and demand for seafood." Rose says the wild fisheries of the world cannot support the future demand with the global population continuing to grow and fish stocks continuing to decline. But on a provincial scale, Rose hopes the fishing industry of the future will be a balance of aquaculture and a wild fishery. "If going forward we can have a balanced approach I think that will be the way for Newfoundland to go into the future."

In summary, the Atlantic media in our sample generally provide more positive coverage of aquaculture than the Pacific or national media, giving more coverage of economic benefits and less coverage of ecological risks. On the other hand, they also give substantially more coverage to economic risks, particularly the actual and potential conflicts between aquaculture and traditional coastal activities and economies.

Conclusions

The media are powerful actors in the aquaculture controversy in Canada. Media organizations play a central role in the dissemination of facts and arguments about the industry and the controversy, and this means that interested actors, particularly opposition groups and industry supporters, invest substantial resources and effort in communicating to and through the media. Our study shows that these efforts are by and large successful: the media are far more likely to quote members of these groups than any other sources. At the same time, however, journalists, editors, and media organizations have their own motives and methods. Aquaculture is a complex topic, with many possible angles and storylines. Journalists must navigate their multiple roles – investigation, explanation, and narration – in telling aquaculture stories. Despite the heavy reliance on sources to write about this issue, our analysis shows that media coverage is still largely driven by journalistic choices, particularly about how to combine themes and voices into narratives.

Our study attempts to find patterns in these choices and examine their implications for the knowledge battlefield over aquaculture in Canada. Although different journalists, editors, and media institutions will make different decisions about aquaculture coverage, our analysis shows that there are clear patterns in these decisions that affect how aquaculture is presented to the public. Overall, the study yielded three significant findings, two of which were unexpected.

The first significant finding is that, as expected, the aquaculture controversy is being reported on from an institutional perspective. The most cited sources in aquaculture coverage are industry members, opposition groups, politicians, and government employees. These are the voices that define and interpret the issues. It is also significant that these voices speak the language of science and expertise (with the notable exception of politicians). Thus, the public discourse about aquaculture is being dominated by a particular set of voices using a particular form of claims making. At the same time, we found that coastal residents are quoted the least in aquaculture coverage. Only 8 percent and 10 percent of articles, respectively, contained a quote from First Nations and non–First Nations coastal stakeholders. We also found that coastal voices appear most often in longer feature or opinion articles, and much less often in standard news articles. This suggests that local voices enter the media discussion of aquaculture only when journalists want to "dig deeper" into the issue.

The second significant finding involves the mutual aversion between economic and environmental themes. Industrial controversies are often presented in the media as a struggle between economic and environmental values (Goodstein 1999). We can see some evidence of this in aquaculture coverage: the two most prominent themes in our media sample are ecological risk or harm (mentioned in 65 percent of all articles) and economic benefits (mentioned in 40 percent of all articles). Our cross-tabulations and *phi* analysis indicate, however, that these themes tend to be mutually averse – that is, they appear together less often than expected given the popularity of each. This leads us to conclude that the media are giving significant attention to both themes, but as part of different narratives. They are stressing either the ecological risks or the economic benefits of aquaculture, and much more rarely presenting them as a "one contra the other." This is in direct contrast to other themes, notably health, where statements about the risks and benefits of aquaculture often go head to head. The aversion is even more puzzling given the tendency for industry and opposition voices to appear in the same articles.

This relates to our third significant finding: the dominance and vulnerability of industry voices in media coverage of aquaculture. As we have seen, industry representatives are quoted most broadly in articles relating to aquaculture. Industry voices appear in 42 percent of all articles, compared with 34 percent for voices from opposition groups. On the surface, it would appear that industry voices are the primary definers of the issue, articulating their perspectives directly to the public through the media. Our study found, however, significant thematic differences between articles where industry voices appear alone and those where they appear in conjunction with other voices. When they appear alone, industry representatives talk about the benefits of aquaculture, particularly the economic benefits of the industry. But when industry voices coincide with opposition or Aboriginal voices in particular, the accompanying themes are overwhelmingly negative. In these cases, industry representatives talk more about risks and harms than about benefits, indicating that they are being brought into the discussions chiefly in a rebuttal role. This is hardly a position of dominance, and may explain in part the extremely negative views of the media held by aquaculture supporters, as described in Chapter 4. Although we do not find evidence of systematic media bias in aquaculture reporting, we do find evidence of selectivity in how the media present voices and themes to tell stories about this industry.

Part 3
Political Economy

Thus far, we have examined how the aquaculture controversy in Canada is intertwined with global and local challenges (Part 1) and with the changing politics of knowledge (Part 2). In Part 3, we look at how aquaculture relates to on-the-ground issues of economy and governance. Federal and provincial governments in Canada have invested heavily in aquaculture development, largely in the hopes that the industry can revive moribund coastal economies, mitigate employment problems resulting from the decline of traditional resource sectors (particularly fisheries and forestry), and re-establish Canada as a major player in global commodities markets (Bastien 2004). As we shall see, however, economic and governance problems have complicated this vision.

Chapter 6 examines how aquaculture fits into the local economies of coastal communities in Canada that have now suffered through nearly two decades of economic turbulence and contraction. As we saw in Chapter 2, local opinions differ on the effects of aquaculture development on communities. Using data from several surveys of aquaculture firms, workers, and communities, we attempt to quantify some of the key impacts of the industry on community development (including employment, job satisfaction, and relations to other local businesses).

Chapter 7 looks at the chronic problems that have plagued aquaculture governance, as well as some of the controversial solutions with which governments are currently experimenting. Aquaculture is a new industry, and aquaculture policy is relatively fluid and open to change. Some new policies are highly innovative and are aimed at broadening the governance tent to include nongovernmental actors – particularly firms, industry associations, and stakeholders – in the formulation of aquaculture policy and regulations. We argue that these initiatives are promising even as they stir up the controversy. Whereas some critics see them as a step towards deregulation, advocates argue that they are a way to help the industry's bottom line while at the same time enhancing its legitimacy and improving environmental compliance.

6
Aquaculture and Community Development

In his 2004 report *Recommendations for Change*, Yves Bastien, the government of Canada's Commissioner for Aquaculture Development, wrote (3):

> Today, Canada ranks twenty-second among aquaculture producing nations and accounts for less than one-third of 1 percent of global farmed fish and seafood production. This level of production falls far below Canada's capacity and capability. Canada has the natural resource base [to] enable it to be in the top three global competitors in aquaculture production ... However, in Canada's drive to become a leading supplier of fish and seafood, we are also aware of the broader objectives of Canadian society to bring economic development to our coastal and rural regions [that is] in harmony with the social fabric of these communities.

This quotation captures the dilemma facing federal and provincial governments in Canada regarding the aquaculture industry. On the one hand, this is an industry with great potential for expansion. Canada is home to vast and varied coastlines, and, as we saw in Chapter 1, salmon aquaculture in particular faces global market pressures that make expansion an urgent priority. On the other hand, "harmony with the social fabric" is equally important. Canada's coastal regions have now endured decades of conflict over resource management and economic development (cf. Hayter 2003; Ommer 2007). These are not idle forces: conflict can and frequently does kill development initiatives that fail to address local social and environmental concerns (Gleditsch 1997; Ramsay 2005).

In this chapter, we investigate how aquaculture fits into local economies and the broader project of coastal development. Not surprisingly, there are strong differences of opinion regarding its impact on coastal communities and economies. To again quote Bastien (2004, 14):

By its very nature, aquaculture exists principally in remote coastal communities, many of which have suffered due to the demise of traditional
fisheries and now have few opportunities for sustainable economic development. Aquaculture occurs in those very areas most in need – areas where
other industries have difficulty taking root and remaining competitive.
Aquaculture has already had a considerable positive impact, leading to the
revitalization of many such communities and providing hope where there
was despair.

Very different conclusions are reached by Marshall, however, in her analysis
of the impacts of aquaculture on traditional economies and practices on
Grand Manan Island, New Brunswick (2001, 350):

> Aquaculture transforms the very meaning of community, and the basis for
> individual and collective identities. The introduction of new productive
> relations that are directed from outside the community, and the alienation
> of the very marine spaces that have sustained the community over genera-
> tions, together threaten to, at the very least, completely transform all social
> relations. If it is to retain any semblance of being "an embedded fishing
> community" ... Grand Manan must be allowed to balance the intrusive
> forces of globalized aquaculture with the traditional bases of cultural liveli-
> hoods and community values.

Like any form of development, aquaculture creates winners and losers
(Wroblewski et al. 2006). The purpose of this chapter, therefore, is to examine
how it impacts already-changing local economies, both positively and negatively (cf. Phyne and Mansilla 2003). This means looking at the employment
impact of aquaculture: how many jobs does the aquaculture industry create,
and are they "good jobs"? It also means investigating the impact of aquaculture on other businesses at the local level. The aquaculture industry is the
"new kid on the block" in many coastal communities, but it is a big kid. It
is clear that aquaculture is disruptive, but it is less clear to what degree, and
in what direction.

To answer these questions, we draw on findings from several research
initiatives. First, we present data from a survey of aquaculture companies
operating across Canada that was conducted in 2003-04 with the cooperation
of the Canadian Aquaculture Industry Alliance (CAIA). The survey was
intended to be a census of all firms involved in the production of aquacultured finfish and shellfish in ocean or fresh water (including hatcheries,
grow-out operations, processors, and firms involved in research and development). It provides a detailed portrait of the structure of employment in
finfish and shellfish aquaculture in Canada. The second study involves a
survey of aquaculture workers conducted in 2004-05 at three major salmon

aquaculture firms and six smaller shellfish aquaculture companies.[1] It provides important insights into what employees of aquaculture firms think of their jobs and their employers. The third research initiative is a case study of the local economy of Port Hardy, British Columbia. Located on the northern tip of Vancouver Island, this community is an aquaculture centre that has also been strongly affected by the decline of traditional resource industries. The case study is based on a survey conducted in 2007-08 of all businesses operating in the community, which enables us to examine the positive and negative impacts of the aquaculture industry on Port Hardy's local economy.

Towards "Post-Resource" Community Development

The Canadian economy has been resource-dependent for a long time (Innis 1956). As recently as 2007, seven of Canada's ten most important export industries were resource-based (Industry Canada 2008).[2] Despite this, resource-dependent communities are in serious decline in many parts of the country. There are many reasons for this, including heightened global competition in commodities markets, local environmental exhaustion, technological innovations that have reduced labour inputs, and policy changes that have "liberated" resource companies from ties to specific regions and communities (see Chapter 1). Although the future of these places is unclear, there is broad consensus that economic diversification is the only way forward for communities that find themselves suddenly as "places without purpose" in a changing economy (Young 2006a; see also Hayter 2000; Halseth and Halseth 2004; Sinclair and Ommer 2006).

Diversification is not easily accomplished, however. In a study of Newfoundland outport communities, Sinclair et al. (1999) found widespread despondency rather than entrepreneurial vim. Young (2006a) found a great deal of confusion among members of several British Columbia coastal communities about what to do next following the decline of the fishery and forest industries. Some in these communities reacted by turning inward, arguing that their communities need to become more self-sufficient and less involved in the broader economy. Alternatively, many community leaders and entrepreneurs were ambitiously imagining new futures for their communities, but in many cases expressed difficulty translating these into concrete economic plans.

Ultimately, diversification means going beyond the "old" resource economy (Apedaile 2004; Howlett and Brownsey 2008). On the ground, we see this happening in two ways. First, diversification happens because it is forced. In many communities, declines in traditional sectors are compelling businesses to either close or become creative in finding new clientele (Young 2006b). Many people are faced with a similarly stark choice: to move away or try to start over in a difficult labour market (Matthews et al. 2009). We label this "negative diversification" because it is due to economic hardship rather than growth. The second form of diversification is more positive. In many (but not all) rural regions, service industries such as tourism have grown substantially and become significant contributors to local economies (Bryden and Bollman 2000). There is also evidence that information technologies are fuelling expansion in retail and business services in rural communities (Apedaile 2004; Young 2009). Finally, governments have become key actors in this regard, actively promoting local diversification through programs such as the Community Futures Development network, which provides loans to entrepreneurs and small businesses in rural and remote areas. By choice or by necessity, many coastal communities are moving towards greater economic diversification.

How does aquaculture fit in with changing local economies, and particularly with the project of economic diversification? According to industry supporters, aquaculture is particularly valuable because it is capable of providing year-round employment (unlike tourism) and because of its multiplier effects across local and regional economies. According to the New Brunswick Salmon Growers Association (NBSGA n.d.a):

> The salmon farming industry has contributed greatly to an economic and cultural revitalization of New Brunswick ... creating employment throughout the province in feed manufacturing, packaging, transportation, supply industries as well as spin-off jobs in the retail sector ... A large number of the [local] salmon-farming related jobs have been the result of the initiative shown by residents in creating a wide range of support businesses.

Opponents argue, however, that aquaculture can have a negative impact on local economies and economic diversification (e.g., Marshall 2003; Cox 2004). For instance, a publication sponsored by the Canadian Centre for Policy Alternatives entitled *Fishy Business: The Economics of Salmon Farming in BC* compares the gross domestic product, employment, and export value of aquaculture with those of the industries that it allegedly threatens – commercial fisheries, the marine sport fishery, and coastal tourism. The report concludes that the latter industries are much more economically important to coastal communities and the province of British Columbia: "There are far more British Columbians – many of whom live in rural regions of the province – benefitting from wild fish stocks than those who benefit from salmon farming" (Marshall 2003, 17). The report uses particularly strong language to describe conflicts between aquaculture and tourism (25):

> Salmon farming may deter tourists if the farms take away from the "natural" or "wilderness" experience for which many come to BC. Some tourist operators are already experiencing the impacts of salmon farming. In fact, there

are direct conflicts between marine ecotourism guides – kayaking and wildlife viewing operations – and the salmon farming industry ... Bays and beaches that were once accessible to the guiding community are now inhabited by salmon farm operations ... The expansion – even the existence – of fish farms is fundamentally at odds with the values that many of BC's tourists hold dear and their reasons for travelling to this province. A choice may have to be made between fish farms and tourism, the latter [being] a larger industry.

In short, aquaculture supporters and opponents present very different portraits of the industry and its role in community development and economic diversification. On the one hand, supporters perceive it as a creator of good jobs and a spur for local entrepreneurialism. On the other hand, critics see it as disruptive and damaging to traditional and emerging industries. In the rest of this chapter, we will present research that shows a more complicated relationship between the aquaculture industry and coastal communities.

The Employment Question: Good Jobs, Bad Jobs, How Many Jobs?

Without doubt, coastal regions in Canada are in the midst of an employment crisis. Unemployment has hovered between 10 and 20 percent since the late 1990s in both Atlantic and Pacific regions, and labour force participation rates are well below the national average (Akyeampong 2007; Sinclair and Ommer 2006, 26), which means that employment is an important factor in evaluating the impact of aquaculture on coastal economies. Estimates of the employment impact of aquaculture vary widely, however. According to the Office of the Commissioner for Aquaculture Development (OCAD), by 2001, the aquaculture industry had created 12,390 full-time, part-time, and seasonal jobs across the country (Bastien 2004, 15). Of these, 8,547 jobs were within the aquaculture industry itself, while an additional 3,843 jobs had been created in related industries such as "boat-building, net and cage manufacturing, and machine shops." These figures circulate regularly in industry publications (e.g., CAIA 2006a). A more recent Department of Fisheries and Oceans (DFO) report, however, estimates direct employment in the aquaculture industry at only 3,900 full-time, part-time, and seasonal jobs (DFO 2005a, 8). Statistical disparities exist at the provincial level as well. The BC Salmon Farmers Association estimates that "about 2,800 people are directly employed in this industry" in British Columbia (BCSFA 2005d), but the government of British Columbia puts the number at 1,900 persons directly employed in 2002 (BC Statistics 2002, 81). Similarly, the NBSGA (n.d.a) claims that aquaculture has created 4,000 direct and indirect jobs in New Brunswick, whereas the DFO report cites 1,250 direct jobs in that province.

The Employment Impact of Aquaculture in Canada

This confusion clouds discussions and debates about the economic impacts of aquaculture. In an attempt to clarify the employment picture, in 2003-04 we conducted a mailed survey of all direct aquaculture firms operating in Canada. A "direct aquaculture firm" is a company involved in some way in the production of aquacultured finfish or shellfish in marine or freshwater environments, including hatcheries, grow-out operations, processors of aquaculture products, and firms engaged in aquaculture research and development.[3] There were 275 responses, for an estimated response rate of 61 percent.[4]

Participating firms were asked to provide information regarding employment across three time periods: (1) last week, (2) during an average week in the preceding year, and (3) during their busiest week in the preceding year. The data from last week are important because they provide an actual statistical snapshot of employment based on real records. It is important to note, however, that the survey was first distributed in February 2003, which, in some aquaculture sectors, is a seasonal low point. Thus, the questions regarding average week and busiest week give some indication of seasonal variation in employment. Findings from these three time periods are shown in Table 6.1.

Several trends are evident. First, these data show that large companies are driving employment. While the categories fluctuate across the time periods, firms with fifty or more employees account for most of the employment in each time period. It is often argued that small businesses drive job creation (e.g., Birch 1987), but this does not appear to be the case in the aquaculture industry.

Second, aquaculture employment varies seasonally. The midwinter last-week employment figures are 12 percent lower than during an average week and 25 percent lower than during the busiest week. This seasonal variation is not evenly distributed. For example, firms with 100 or more employees show comparatively little variation. The difference between last week and an average week for these large firms was only 42 persons, or 2.0 percent, which suggests that they have a stable labour force that is not notably affected by seasonal downturns.[5] These firms do expand in their busiest week, however, reporting 398 more persons employed, a 19.5 percent increase over the last week figures. In contrast, smaller firms fluctuate significantly. For instance, 98 firms reported that they had "zero employees" during the last week. This number drops to 61 and 62 during an average week and the busiest week, respectively. This indicates that over one-third of small firms hire at least one additional employee at some point in the year. We can assume that this "category jump" is occurring among firms with 1-4 employees as well. Although the latter category appears to be stable across the time periods (with 103 firms, 111 firms, and 96 firms for last week, average week, and

Table 6.1

Total number of employees in aquaculture firms that participated in the study, by company size

Company size	Last week			Average week			Busiest week		
	Total employment	Mean number of employees per firm	Number of firms	Total employment	Mean number of employees per firm	Number of firms	Total employment	Mean number of employees per firm	Number of firms
0 employees	0	–	98	0	–	61	0	–	62
1-4 employees	246	2.39	103	263	2.37	111	236	2.44	96
5-24 employees	463	9.65	48	679	9.30	73	794	9.45	84
25-49 employees	352	35.20	10	541	36.07	15	549	36.60	15
50-99 employees	366	73.20	5	323	80.75	4	343	68.80	5
100+ employees	2,090	190.00	11	2,132	193.82	11	2,488	191.38	13
Total	3,517	12.79	275	3,938	14.32	275	4,410	16.04	275

busiest week, respectively), the next greater category changes significantly. Specifically, the number of firms with 5-24 employees climbs from 48 during the last week to 73 during an average week, to 84 during the busiest week. This suggests that the most seasonal fluctuation in employment is occurring among small to mid-sized businesses.

Our next task is to estimate the overall employment impact of aquaculture in Canada. In order to do this, we need to extrapolate our data using information about the whole population. This information came from CAIA, which examined our sample and identified 6 firms as likely to have 100 or more employees, 6 firms as likely to have 50-99 employees, 18 firms with 25-49 employees, 127 with 5-24 employees, and the rest as likely to have 4 or fewer employees. Although these estimates are open to error, they enable us to estimate total employment within each size category, which we then sum to get an overall estimate. The estimated employment figures are given in Table 6.2. These range from a low of 5,148 persons employed last week to 5,237 employed in an average week, to a high of 5,565 persons employed in the busiest week.[6]

In our opinion, these figures are a highly reliable estimate of total direct employment in aquaculture in Canada at the time of the survey (2003). Given that these calculations are based on estimates of the number of firms in each category, we can conservatively add or subtract 10 percent from each of the figures to account for error, and state with considerable confidence that total employment in direct aquaculture firms in Canada in 2003-04 ranged from 5,000 to 6,000 employees, even taking into account seasonal adjustments.

This estimate is significantly lower than the 8,500 direct jobs estimated for 2001 by OCAD (Bastien 2004, 15), but is higher than DFO's estimate of 3,900 direct jobs (DFO 2005a). Our provincial breakdowns also differ from some published figures. For instance, our estimate of total direct employment in aquaculture in New Brunswick is between 1,710 and 1,850 jobs, which is lower than the NBSGA's estimate of 4,000 direct and indirect jobs in salmon

Table 6.2

Estimated total employment in direct aquaculture firms in Canada

Company size	Last week	Average week	Busiest week
0-4 employees	710	704	725
5-24 employees	1,226	1,181	1,200
25-49 employees	493	541	549
50-99 employees	439	485	412
100+ employees	2,280	2,326	2,679
Total	5,148	5,237	5,565

farming alone (it is unclear how indirect jobs were calculated in their analysis) (NBSGA n.d.a). Similarly, our estimate for British Columbia is between 1,880 and 2,000 direct jobs in aquaculture (shellfish and finfish), which is in line with the BC Statistics estimate of 1,900 direct jobs (BC Statistics 2002) but is substantially lower than the BCSFA's estimate (2005d) of 2,800 direct jobs in salmon aquaculture alone.[7]

Next, we examine the structure of employment in Canadian aquaculture. Thus far, we have looked at the number of jobs in the industry without addressing sectoral differences (for instance, between finfish and shellfish aquaculture) or differences in types of employment. A sectoral breakdown of employment among responding firms for the last-week period is shown in Table 6.3. This table confirms the dominance of the marine finfish sector (salmon aquaculture) in terms of employment. Although only 10 percent of firms responding to the survey indicated that they were involved in this sector, it accounts for nearly 50 percent of total employment in the industry. In contrast, nearly a third of firms indicated involvement in shellfish aquaculture but the average number of jobs per firm in this sector is significantly lower (11 as opposed to 48). A surprisingly high number of firms indicated involvement in freshwater finfish aquaculture (typically trout farming): 58 of them, representing 21 percent of all respondents; however, the employment impact of freshwater aquaculture, with an average of 6.5 employees per firm, is significantly less than that of the ocean-based sectors.

Table 6.4 provides information on the types of jobs in the aquaculture industry, specifically what proportion are full-time, part-time, and seasonal or casual jobs. These figures show that most employees in the aquaculture industry are engaged in *full-time* work. Over 80 percent of all workers during the last-week period were engaged full-time. Again, it appears that the larger firms are driving this trend, although all types of firm tend to employ full-time workers. Nearly half of all companies responding to the survey

Table 6.3

Employment patterns by aquaculture industry sector: last week

	Employment		Percentage of responding firms indicating involvement	Average number of
Sector	N	%	in the sector	jobs per firm
Shellfish	943	32.6	32.4	10.6
Marine finfish	1,340	46.4	10.2	47.9
Freshwater finfish	379	13.1	21.1	6.5
Other	229	7.9	8.7	9.6
Total	2,891	100.0		

indicated that they employed at least one person full-time during the last week, while only one-third indicated employing a part-time employee, and one-quarter a seasonal or casual employee.

The trends for busiest-week employment differ slightly. During this period, we see a significant expansion in seasonal and casual employment. Whereas only 9.6 percent of workers fall under this category for the last-week period, during the busiest week seasonal or casual employees constitute 18.2 percent of the total workforce. Nearly half of all firms reported hiring casual workers during peak times, up from one-quarter during the last week. Nevertheless, full-time employment is predominant even during the busiest week, accounting for 72 percent of all jobs.

We turn now to the question of who is working in aquaculture. As shown in Table 6.5, nearly half of all employees in the aquaculture industry are young adults (twenty-one to thirty-five years of age). This is a significant finding, because young people are particularly vulnerable to economic change in rural communities (Cartmel and Furlong 2000). The exodus of young people from small towns is a longstanding social problem that is exacerbated during times of economic stress (Jackson et al. 2006). If aquaculture is providing stable, full-time employment to young people in rural regions (as our figures indicate), then this is a major social and economic benefit to communities struggling to retain and attract youth. Moreover, the last column in Table 6.5 gives the increase or decrease in the number of employees from last week to busiest week, thus showing that aquaculture firms are hiring teenagers as part of their seasonal expansion. The teenaged workforce increases 60 percent from the last-week to the busiest-week period, suggesting that opportunities are being made available for high school students to work seasonally in the industry.

The gender of aquaculture employees is shown in Table 6.6. Nearly three-quarters of aquaculture workers in responding firms are male. This imbalance may be due in part to the traditional gender skew in many of the occupations related to aquaculture, including diving (to clean and repair nets and other equipment) and transportation. Moreover, many aquaculture sites are remote and require long stays on the water. This certainly does not explain or excuse the absence of women, but the long-stay work arrangement has an extensive history in male-dominated rural and resource activities, from the logging camp to the offshore fishing vessel. On the other hand, there are many jobs in the aquaculture industry that should not show a gender imbalance, including work in administration, veterinary services, husbandry (hatcheries), and processing. Overall, the under-representation of women in the aquaculture workforce is troubling. Table 6.6 also shows that, unlike young people, women are not a "seasonal reserve." More women are employed at peak times but the proportion of female employees does not differ substantially from the last-week period.

Table 6.4

Full-time, part-time, and seasonal/casual jobs in the aquaculture industry

	Last week				Busiest week			
	Employment		Percentage of firms employing	Average number of jobs per firm	Employment		Percentage of firms employing	Average number of jobs per firm
Type of job	N	%			N	%		
Full-time	2,870	81.2	44.7	10.44	3,187	72.2	49.1	11.59
Part-time	324	9.1	33.1	1.18	422	9.6	34.2	1.53
Seasonal/casual	342	9.7	24.0	1.24	805	18.2	51.6	2.93
Total	3,536	100.0		12.86	4,414	100.0		16.05

Table 6.5

Age of employees in the aquaculture industry: last week compared with busiest week

	Employment		Percentage of firms employing	Busiest week	
Employee age	N	%			
20 and younger	264	7.4	17.5	+159	(+60%)
21 to 35	1,610	45.5	45.1	+387	(+19%)
36 to 50	1,227	34.7	45.5	+242	(+20%)
51 and older	439	12.4	33.8	−13	(−3%)
Total	3,540	100.0			

Table 6.6

Gender of employees in the aquaculture industry: last week compared with busiest week

Employee gender	Employment		Percentage of firms employing	Busiest week
	N	%		
Male	2,543	71.9	63.3	+609 (+23 %)
Female	993	28.1	40.0	+269 (+27 %)
Total	3,536	100.0		

Table 6.7

Aboriginal employees in the aquaculture industry: last week compared with busiest week

Aboriginal employee	Employment		Percentage of firms employing	Busiest week
	N	%		
Full-time	156	73.9	8.0	−2 (−1%)
Part-time	22	10.4	4.7	−3 (−14%)
Seasonal/casual	33	15.6	2.9	+47 (+143%)
Total	211	100.0	10.5	+40 (+19%)

The final demographic criterion considered is Aboriginal employment. Both the Atlantic and Pacific coasts of Canada are home to Aboriginal groups that have long traditions in coastal industries. As shown in Table 6.7, however, Aboriginal employment in aquaculture is low. Only 11 percent of firms responding to our survey reported having any Aboriginal employees, and Aboriginal persons constitute only 6 percent of the total workforce reported. It is possible that some aquaculture firms operated by Aboriginal groups chose not to complete the survey. Nevertheless, the lack of widespread Aboriginal employment should be of concern to the industry. As we saw in Chapter 2, some firms are actively pursuing better relationships with Aboriginal groups. These employment figures suggest that efforts to reach out may be limited to a few firms in a few places. On a more positive note, we see that most Aboriginal employees are engaged in full-time work (although significant numbers are engaged seasonally).

Good Jobs, Bad Jobs?
The quality of aquaculture jobs is difficult to measure. Work in the aquaculture industry often involves long hours, physical labour, and time away from

family. As mentioned earlier, however, these demands are not uncommon in rural and resource industries such as forestry and fisheries. Indeed, many have assumed that displaced resource workers would be attracted to aquaculture precisely because of "the lifestyle in common" (e.g., Stickney and McVey 2002). At the same time, it is often argued that because aquaculture means "going to work for a company" – often a large, centralized multinational firm – it is inherently less gratifying than traditional work in small-scale owner-operated fisheries (e.g., Wurts 2000).[8]

In this section, we look at the quality of aquaculture jobs across two dimensions. First, we use the survey of aquaculture firms to look into the structure of incomes and benefits available to workers. We have already seen that most jobs in the aquaculture industry are full-time and relatively stable, but we have not yet examined issues of compensation and of family security (in the form of benefits). Second, we will look at indicators of job satisfaction from a survey of aquaculture workers themselves. Although that survey was performed on a smaller scale, it gives us a window into how aquaculture employees think about their work.

The distribution of annual earnings among aquaculture employees is shown in Table 6.8 (these figures come from our survey of firms). The modal category is income between $25,000 and $35,000, and the second-largest category is $15,000 to $25,000. Overall, 78 percent of aquaculture workers made under $35,000 per year in 2003. These income figures are low. The mean individual income for working people in Canada according to the 2001 census was $31,757 (Statistics Canada 2003). In contrast, according to our survey the average annual income among aquaculture employees was approximately $28,300 in 2003.[9] Aquaculture, however, takes place primarily in rural, coastal communities that have substantially lower average incomes than in Canadian cities. This is particularly true in Atlantic Canada, where, for instance, the annual average income among employed persons in Grand Manan, New Brunswick, was $23,500 in 2001. For Malpeque Bay, Prince Edward Island, the figure is $19,000; Bay D'espoir, Newfoundland and Labrador, $22,000; and Les Îles-de-la-Madeleine, Quebec, $23,000.[10] The average aquaculture wage is high in these contexts, although not overly so. The situation is different on the Pacific coast, which has traditionally enjoyed higher wages than the Atlantic coastal region (Sinclair and Ommer 2006, 25). The heart of the aquaculture industry in British Columbia lies in places such as Campbell River (average income among employed persons in 2001, $29,500), Tofino ($28,000), and Port Hardy ($30,500). Here, aquaculture wages are slightly below the average among working persons. Aquaculture is also, however, a major contributor in smaller and predominantly Aboriginal communities such as Klemtu ($17,800) and Ahousaht ($21,500), where incomes tend to be low.

Table 6.8

Income distribution for aquaculture employees

Income level	Number of employees	Percentage of employees
Under $15,000	558	18.4
$15,000-24,999	822	27.2
$25,000-34,999	973	32.2
$35,000-44,999	313	10.4
$45,000-54,999	185	6.1
$55,000-64,999	87	2.9
$65,000-74,999	40	1.3
Over $75,000	45	1.5
Total	3,023	100.0

We also find significant differences in income between shellfish and marine finfish aquaculture. The average wage in finfish aquaculture is approximately $31,296, compared with only $24,633 in shellfish aquaculture.[11] Put another way, 65 percent of workers in finfish aquaculture make over $25,000, compared with only 43 percent of shellfish workers. The discrepancy increases among higher incomes: whereas 30 percent of finfish aquaculture workers earn $35,000 or more, only 13 percent of shellfish workers do the same.

So, are these jobs good or bad? The answer appears to depend on context and point of view. On the one hand, finfish aquaculture is often criticized as being exploitative (Wurts 2000), with major multinational firms said to be paying minimal wages for maximum output (Cox 2004, 33). Our survey finds, however, that average wages in *finfish aquaculture* are almost exactly equal to the national earnings average (around $32,000). On the other hand, shellfish aquaculture is often applauded as "offering significant economic benefits to [communities] while being compatible with traditional values" (Tollefson and Scott 2006, 7), but our research shows that earnings in this sector are considerably lower.

A second structural measure of job quality involves employee benefits. The survey asked respondents to indicate which benefits they provided "for the majority of their full-time employees." The results are shown in Table 6.9, which indicates that many aquaculture workers are poorly served in terms of benefits. A paid vacation (which is a legal requirement for full-time employees) is the only employer benefit provided by most firms in either shellfish or finfish production. Pensions are rare, as are paid sick leave, life insurance, disability insurance, and insurance against accident. Medical and dental benefits are provided by only a small percentage of shellfish firms, and less than half of finfish firms.

Table 6.9

Percentage of aquaculture firms providing employee benefits, by sector

Type of benefit	Firms with a shellfish grow-out (% of total firms [70])	Firms with a finfish grow-out (% of total firms [54])
Basic benefits		
Paid vacation	73	87
Paid sick leave	17	32
Insurance and safety		
Life insurance	7	44
Disability insurance	17	41
Accidental death/dismemberment	14	44
Employee and family assistance	7	19
Services		
Dental plan	14	48
Supplemental medical insurance	16	54
Paid parental leave	4	4
Pension	4	30
Childcare benefits	0	4
Employee expenses		
Housing allowance	10	19
Education reimbursement	19	39
Automobile allowance	10	17

Overall, employees of finfish-producing companies have more comprehensive benefits than employees in the shellfish sector, but this appears to be largely a function of the size of firms. For instance, among the smallest firms (with one to four employees), only 12 percent indicated that they provide a dental plan, 12 percent life insurance, and 16 percent medical insurance. These benefits are much more common among mid-sized firms (employing between twenty-five and forty-nine persons), with 67 percent offering a dental plan, 67 percent life insurance, and 67 percent medical insurance. Among the largest companies (with fifty or more employees), the figures are 94, 88, and 88 percent, respectively. On average, shellfish aquaculture firms tend to be smaller (see Table 6.3), and are therefore disproportionately among those firms that are either unwilling or unable to offer comprehensive employee benefits.

All of the data presented thus far lead us to conclude that there are two faces to employment in the Canadian aquaculture industry. On the one hand, there is strong evidence that employees of large aquaculture firms enjoy relative job security, year-round full-time employment, a reasonable living wage, and access to important employee benefits. These "good jobs" tend to be with larger firms that are engaged in marine finfish aquaculture, which in Canada is overwhelmingly salmon aquaculture. As we have seen throughout this book, salmon farming is the most controversial activity in Canadian aquaculture, but this is also where the employment impact of aquaculture is most significant. Large companies in this industry employ significantly more people than do smaller firms (see Table 6.2), and this employment is mostly full-time and open to young people in rural communities.

On the other hand, there is evidence that many firms in the aquaculture industry are unable or unwilling to offer job security, high wages, or significant employee benefits. Nearly half of all aquaculture firms engage a contingent workforce (seasonal or casual employment). Although this type of work is often desirable or beneficial, as when firms employ youth for the peak summer period, it can also reflect poor employment conditions. Indeed, we see from Table 6.7 that although Aboriginal persons are under-represented in aquaculture employment generally, they are over-represented in the casual workforce.

Job Satisfaction

We turn now to the issue of job satisfaction among aquaculture workers. As mentioned earlier, our primary source for this analysis is a survey of workers at three salmon aquaculture and six shellfish aquaculture firms. Questionnaires were distributed and collected directly (i.e., without the involvement of management). In receiving permission to conduct this study, we guaranteed anonymity to the firms involved, but we can state that both Atlantic and Pacific regions are represented. The study is somewhat limited in that we were able to conduct only one distribution of questionnaires, which restricted our response rate to 32 percent (N = 132). Although this response rate is lower than we had hoped, it is within the normal range for survey research (Kaplowitz et al. 2004). Despite this limitation, the study offers a rare glimpse into employee attitudes towards their employers, workplace, and industry.

The survey contained multiple questions pertaining to job satisfaction. Most of these involved asking respondents to indicate their level of agreement with statements about their work and workplace.[12] Our findings, which are summarized in Table 6.10, suggest that aquaculture employees have complex views about their work, very positive about some aspects but less enthusiastic about others. For instance, both salmon aquaculture and shellfish aquaculture workers generally report satisfaction with workplace relations. There is strong agreement that "relations among workers at my workplace are good," and moderate to strong agreement that "relations

Table 6.10

Job satisfaction among shellfish and salmon aquaculture workers

Statement	Sector	Agree or strongly agree (%)	Neither agree nor disagree (%)	Disagree or strongly disagree (%)
Relations between manage-	salmon	72	16	12
ment and employees at	shellfish	62	27	8
my workplace are good				
Relations among workers at	salmon	83	11	5
my workplace are good	shellfish	81	15	0
I am satisfied in my job	salmon	70	15	14
	shellfish	54	19	20
My job is secure	salmon	60	22	1
	shellfish	80	4	12
I can see myself working	salmon	58	26	17
in the aquaculture	shellfish	20	31	46
industry as a career				
Given the chance, I would	salmon	34	34	31
change my present type	shellfish	54	23	20
of work for something				
different				
My work is interesting	salmon	72	16	11
	shellfish	39	15	43
The opportunities for	salmon	34	37	29
advancement in my	shellfish	16	38	42
job are good				

between management and employees at my workplace are good," particularly in the finfish sector. These are important indicators of job satisfaction, and they suggest that conflicts in the workplace are rare. Finally, 80 percent of shellfish aquaculture workers and 60 percent of salmon aquaculture workers agreed with the statement "My job is secure," while only 12 percent and 18 percent, respectively, disagreed. Labour research shows that job security is an important component of job satisfaction, particularly among men, less-skilled employees, and primary wage earners supporting a household (Sousa-Poza 2000). In the economically stressed rural coastal environment, this is an important consideration in assessing job quality.

Although employees in salmon and shellfish aquaculture express similar opinions on issues of workplace relations and job security, they diverge on most of the other measures. For instance, only 54 percent of shellfish aquaculture workers agreed with the bellwether statement "I am satisfied in my job," compared with 70 percent of salmon aquaculture workers. Both figures are low compared with most cross-industry studies, which consistently find that job satisfaction among all workers runs at about 80-85 percent (Blanchflower and Oswald 1999), but salmon aquaculture workers are notably more satisfied than shellfish aquaculture workers. A similar divergence is evident for the statement "My work is interesting": 72 percent of salmon aquaculture workers agreed, compared with only 39 percent of shellfish aquaculture workers. Finally, the survey included the statement "I can see myself working in the aquaculture industry as a career." Again, we see a split between salmon (58 percent agreement) and shellfish aquaculture (only 20 percent agreement). Perhaps more telling, 46 percent of shellfish aquaculture workers disagreed with this statement, suggesting that for many workers, a long-term career in shellfish aquaculture is undesirable. Indeed, 54 percent of shellfish aquaculture workers agreed that "given the chance, I would change my present type of work for something different." This is a key measure of dissatisfaction, and its high frequency among shellfish workers is suggestive (in contrast, only 34 percent of salmon workers agreed).

Salmon and shellfish aquaculture workers also report different experiences with on-the-job skills. One of the most commonly articulated complaints against technologically intense industries is that they reduce both the need for labour and the discretion of labourers. This is the "de-skilling hypothesis," wherein more technology in the workplace (both in the form of tangible technologies and in the form of management "techniques") is assumed to reduce the need for human skills and judgment (Braverman 1974; Sennett 1998). Salmon aquaculture in particular has become technologically intensive in recent years, with the partial or full automation of tasks such as feedings, site monitoring, and fish health. This has put downward pressure on the number of people employed in salmon farming relative to output (BC Statistics 2002, 35). It is unclear, however, whether this has resulted in the "de-skilling" of the remaining labour force.

The de-skilling hypothesis is controversial, with detractors pointing to many examples where the presence of technology in the workplace has a "skilling" effect that in turn enhances job satisfaction (e.g., Attewell 1987). Our survey of aquaculture workers asked respondents: "Since you began working in your current job, has the level of skill required to perform your work increased, decreased, or stayed the same?" Sixty-two percent of salmon aquaculture workers indicated that the level of skill required for their job had *increased*. Only 1 percent indicated a decrease in skills required, while

38 percent stated that there was no change. Again, these opinions contrast with those held by shellfish aquaculture workers. While the latter agreed that there had been little de-skilling in their work (in fact, there were no claims of skill decrease), only 30 percent indicated an increase in skills and 60 percent indicated no change.

There is also a qualitative difference in how salmon and shellfish workers talk about changes in skills required. The survey asked respondents who indicated a skill change to "briefly describe how the skill level required in your job has increased or decreased." Among salmon aquaculture workers, these discussions typically highlight taking on new tasks or responsibilities:

Feeding fish [is now] more technical. [We're] tracking things that we didn't even know about ten years ago. (Survey reference 385)

I've learned to identify plankton [and] I've learned to identify any problems with the fish. (Survey reference 293)

[There is] more monitoring of fish behaviour. More emphasis on recirculation systems and therefore [this needs] higher skills in efficient system monitoring. [We have] also increased the frequency of sampling for fish health diagnosis. (Survey reference 109)

To do this job now I have to learn every aspect of salmon farming. (Survey reference 86)

In comparison, many of the shellfish workers who indicated a skills increase were in fact referring to "getting better at the job." This often has little to do with learning new skills:

I've learned to shuck [shellfish] with better quality, and now have better speed. (Survey reference 25)

I used to shuck slow – now [I'm] faster. (Survey reference 36)

Now I shuck faster and cleaner. (Survey reference 30)

Although our sample is too small for us to definitively conclude that employee morale is higher in salmon than in shellfish aquaculture, the findings are suggestive. As mentioned earlier, it is often assumed that shellfish aquaculture is more culturally appropriate for small coastal communities and the rural labour force than salmon aquaculture (Shumway et al. 2003; Tollefson and Scott 2006). The survey of workers appears to contradict this assumption, as levels of job satisfaction appear to be high among salmon farming employees (though still lower than those found in most cross-industry studies) and substantially lower among shellfish aquaculture workers.

New Kid on the Block: Aquaculture and Communities

To understand the full impact of aquaculture on rural communities, we need to go beyond questions of direct employment and job quality. In the rest of this chapter, we look at how aquaculture actually fits in with changing community economies. Like any form of development, aquaculture is disruptive: it has an impact on local social and economic dynamics, it affects other businesses, and it even changes the composition of communities (for example, by encouraging some people to move to the community and others to leave it). To examine this, we use data from the survey of aquaculture firms, the survey of aquaculture workers (both of which were considered earlier), and a survey of local businesses operating in Port Hardy, British Columbia. As will be discussed below, Port Hardy is an important centre for the salmon aquaculture industry in the province, being adjacent to the Broughton Archipelago, where many farms are located. At the same time, it is one of the communities that has been hardest hit by restructuring in traditional resource sectors. This makes it an important case for looking at the impact of aquaculture on changing local economies.

Challenges Facing the Aquaculture Industry in Coastal Canada

In order to understand how aquaculture fits in with changing coastal economies, we need to look at the question from two perspectives: that of firms and that of communities.

Aquaculture companies make demands on their host communities. Salmon aquaculture, in particular, is a "continuous-care" operation. Labour is required on a daily and even hourly basis to ensure the survival and continuity of the stock, which means that there are no days off in the life of an active fish farm (shellfish aquaculture is less demanding in this respect). Second, we saw in Chapter 1 that markets for aquaculture products are extremely competitive. For small enterprises in particular, survival means reducing costs and being flexible. On the positive side, these pressures lead to innovation, and over half of small aquaculture businesses (fewer than twenty-five employees) reported that they had expanded into a new product, market, or service within the last five years. On the negative side, firms are under pressure to reduce costs, which exerts downward pressure on wages and leads to labour and recruitment problems. According to our survey, 60 percent of small and medium-sized firms (fewer than twenty-five employees) that had job openings in the previous year reported difficulty in filling positions at the "general labourer" level.

The dependence on labour, along with the need to be flexible and reduce costs, puts strain on aquaculture firms that often translates into ill feeling towards communities and the local labour force. This is captured in an open-ended question included in the survey of firms: "What do you think are the biggest challenges in recruiting and retaining employees?" Many firms identify the need for continuous labour as a problem:

The isolation and work environment are not for everyone. The rotations in and out of the sites are hard on families and relationships. The work itself can alternate from hard physical labour to periods of sheer boredom watching underwater camera images. (Salmon firm: Respondent 86)

A lot of the work is night time work, and people are not interested in this. Piece work is a major factor – many do not like this. Both combine to make it hard to keep and attract employees. (Shellfish firm: Respondent 1)

The work is hard, long and labour intensive, and dirty. Due to [this] ... it is sometimes difficult to find young workers. (Shellfish firm: Respondent 216)

Another common theme was the cost-reduction dilemma. Many smaller ventures in particular explicitly state that they are unable to pay the wages that they would like to pay:

The problem with recruiting and retaining employees is that my company is not big enough to hire these people for any extended period of time, financially speaking. (Trout firm: Respondent 44)

Finding people willing to pick for $10.00 an hour (all we can afford) is a real problem. (Shellfish firm: Respondent 167)

The biggest challenge in recruiting and retaining employees is getting a fair market value for produce to ensure full-time wages high enough to support workers and their families. (Shellfish firm: Respondent 57)

Many other firms, however, responded to this question by expressing frustration with the character and behaviour of the local workforce:

The biggest challenge in recruiting employees is finding people with common sense and with motivation. They seem to be very few and far between. (Salmon firm: Respondent 82)

[Finding employees] with a willingness to work 12 months in a row. Enough brains to do the job. (Type of firm unknown: Respondent 78)

The only challenge we have found so far is finding hardworking and dedicated individuals. (Shellfish firm: Respondent 131)

Students are cheaper to hire than mature employees, but it remains a problem to find students with a genuine desire to work. I always feel like I'm inconveniencing them, despite paying over minimum wage. (Trout firm: Respondent 32)

Comments such as these reflect a possible source of friction between the aquaculture industry and host communities. It is possible that many *potential* aquaculture workers either leave the industry or avoid it altogether because of wage and compensation problems. While we saw earlier that job satisfaction is generally high (in the limited number of workplaces surveyed), such a survey does not tell us about potential workers, those who could work in the industry but, for various reasons, do not.

We do not mean to suggest that aquaculture is a bad fit for many communities and workers, but rather that the fit is imperfect. We find further evidence of tensions in other responses to the open-ended "recruiting and retaining employees" question. Specifically, many respondents identified Employment Insurance (EI), formerly known as Unemployment Insurance (UI), as the root of labour difficulties.

We find the workforce around this area is deplorable. They do not want to work by the hour, they only want to work for six months of the year and then draw EI the remaining six months ... EI is too good. If you could stay home and draw 55 percent of your pay and not have to go out in all weather, what would you do? We have had people quit here, go on EI, say that they didn't want to work five days one week and six the next – [that] it was too stressful. They received EI with no problem even though they had quit! (Shellfish firm: Respondent 9)

The largest challenge is the Employment Insurance program ... [It's] hard to find staff who want to work. (Shellfish firm: Respondent 95)

We cannot compete with UI. (Shellfish firm: Respondent 48)

The EI system has to be changed. How to fix the EI system you ask? Don't get me started. (Shellfish firm: Respondent 204)

These complaints against the EI program are not trivial, and are clearly a significant problem for aquaculture firms (particularly smaller firms that have difficulty offering high wages). Again, however, the complaints may reflect a deeper tension between the industry and coastal communities.

Many communities have longstanding work traditions that have evolved as a reflection of coastal living. For instance, many traditional resource activities – including logging and commercial fishing – are seasonal in nature. Coastal families and communities have adapted to this reality by carrying out different jobs and activities at different times of the year (Cadigan 2006), and when state programs such as EI emerged, they became part of these coping and adaptation strategies (Ommer 1999). Aquaculture also varies seasonally, but not to the same degree. Its stability is an important *strength* of aquaculture employment, but it is also important to recognize that this affects how the industry fits into communities.

Our survey of aquaculture workers provides an additional insight into this labour tension. The survey asked respondents about their previous employment history. We found that *very few* aquaculture workers had any background in traditional resource industries. Despite the often-heard claim that aquaculture "provides an ideal occupational alternative for displaced fishermen" (Shumway et al. 2003, 17), only 9 percent of respondents had ever worked in fisheries. Likewise, only 7 percent had ever worked in forestry. The most common previous job for aquaculture workers was in another branch of the aquaculture sector (29 percent of all workers), followed by construction (20 percent) and service industries (20 percent), including retail, hospitality, and odd jobs. While our data about workers are limited in scope, they suggest that workers in traditional resource sectors are not migrating to aquaculture in large numbers. In our sample, there were more former government employees (10 percent) than former participants in commercial fisheries. It may be that aquaculture's year-round work demands are in conflict with the seasonal work traditions that are highly valued by workers in traditional resource industries.

The Impact of Aquaculture on Local Economies
As noted at the beginning of this chapter, the impacts of aquaculture on local economies are disputed. On the one hand, aquaculture boosters argue that the industry is revitalizing moribund coastal economies – not only creating new employment opportunities but also supporting many auxiliary businesses (e.g., Bastien 2004, 16). On the other hand, critics argue that aquaculture is damaging, or at the very least threatening, local economies. From this perspective, aquaculture menaces established industries (such as commercial fisheries) as well as emerging ones (such as tourism). Moreover, critics often argue that industry claims about the direct and indirect economic benefits of aquaculture are exaggerated (Marshall 2003, 16).

It is difficult to conclusively measure the impact of aquaculture on local economies, but several important pieces of research can help us get a rough idea of the positive and negative impacts of the industry. First, we can use "multiplier ratios" to estimate how many indirect jobs are being created by the aquaculture industry in support activities such as transportation, machine works, construction, and so on. Multiplier ratios can be calculated using various methods, but generally relate to the amount of money that firms in different sectors typically spend on non-labour expenditures (Horne 2004, 56; Bivens 2003). In practice, multiplier ratios for aquaculture vary significantly according to different studies. Salmon and Kingzett (2002) report a multiplier ratio of 0.66, or two-thirds of an indirect job for every one direct job. Figures presented by Bastien (2004) are based on a multiplier ratio of 0.72 for salmon aquaculture and 0.16 for shellfish aquaculture. A DFO report uses a ratio of 1.49 (DFO 2005b). When we combine each of these ratios with the direct employment estimates presented earlier in this chapter, we arrive at anything from 3,300 indirect jobs to 8,940 indirect jobs nationwide. Multiplier ratios do not tell us whether or what proportion of these jobs are held in coastal regions, however.

A second measure is provided by our survey of aquaculture firms, which asked responding firms to indicate which activities the company handles internally, contracts out, or both. Fourteen specific areas related to aquaculture were listed, and the findings are presented in Table 6.11. This table shows that contracting out of services and expertise is quite common in both the finfish and shellfish aquaculture sectors. Shellfish firms in particular contract out many services, including processing, diving, research and technical services, transportation, and veterinary services. Finfish firms tend to do more in-house, but also commonly use contracted services in diving, research and technical services, transportation, and veterinary services. Moreover, finfish firms have a tendency to blend in-house work and contracting out, notably for construction, net and cage maintenance, and environmental monitoring. Finally, we note that both small and large firms are heavily involved in contracting (see Matthews 2004, 38). Smaller companies (with fewer than twenty-five employees) tend to contract out processing much more frequently than larger companies, but larger firms are significantly more likely to contract out other core activities, such as diving, construction, maintenance, transportation, and waste management. It appears that smaller firms often try to make do with internal labour for activities such as construction and waste control, whereas larger firms are likely to be in a much better position to use contracts to reduce labour costs.

Overall, these data suggest that the aquaculture industry has a strong impact on other firms, but we still do not know *where* this presence is being felt. From the point of view of community and coastal development, local contracts are infinitely more valuable than contracts with faraway or urban-based firms. Although we do not know whether aquaculture firms are contracting locally, the breakdown of contracted activities suggests that a large proportion of business does remain in coastal communities. For instance, activities such as diving, maintenance, construction, boat services, and

Table 6.11

Services performed internally, contracted out, or both by the shellfish and finfish sectors

Activity	Shellfish (number of firms)			Finfish (number of firms)		
	Internal	Contracted out	Both	Internal	Contracted out	Both
Processing	37	50	6	34	15	2
Diving	46	39	9	13	15	11
Net and cage maintenance	65	2	2	32	5	15
Construction	71	16	15	33	11	24
Research/technical services	42	26	12	18	19	17
Setting of moorings/ anchors	83	11	10	25	4	8
Boat services (e.g., transport products)	68	25	9	17	13	5
Trucking	41	51	12	26	28	12
Therapeutant treatment	8	14	0	36	12	9
Veterinary services	3	19	1	11	41	6
Seed collection	79	28	8	23	10	3
Grow-out	113	2	2	65	2	6
Waste management	66	9	6	41	15	11
Environmental monitoring	60	20	10	36	10	21

environmental monitoring are place-specific. There is no guarantee that these services are provided by local businesses, but at least some local involvement is highly likely.

To examine the actual impact of aquaculture on local economies, we turn to a case study based on in-depth research in the community of Port Hardy in 2007-08. Port Hardy is a community of 3,800 people on the northern tip of Vancouver Island that has become an important centre for the salmon aquaculture industry. Our research there involved an in-person survey of all businesses operating in the community.[13] Overall, 181 businesses participated in the study, for a response rate of 65 percent. It is important to note that in every case, interviews were conducted with either owners or managers of businesses – in other words, with people who have both a deep knowledge of the business and a measure of decision-making authority.

The overall purpose of the study was to investigate how local businesses are coping with and adapting to economic change.[14] Prior to the mid-1990s, Port Hardy was an archetypal resource town, and the local economy was dominated by commercial fisheries, forestry (particularly logging), and a nearby copper mine operated by international giant BHP. This economy was highly prosperous, and the average individual income in Port Hardy in 1996 was over $3,000 above the provincial average. A series of economic crises since that time have redefined the local economy, however. In 1996, the copper mine closed when the deposit was exhausted. At about the same time, the lucrative salmon fishery suffered a serious decline due to fleet reduction and conservation efforts spearheaded by DFO (Brown 2005). Finally, forestry in the region has suffered since the mid-1990s, due in large part to competition from newer and more efficient mills in the Interior of the province (Hayter and Barnes 1997).

These changes have significantly impacted Port Hardy's business community. According to a recent government study, Port Hardy is British Columbia's least economically diversified and most resource-dependent community (Horne 2004). Ironically, the decline in traditional resource economies has made Port Hardy's economy *less diverse* over time (Horne 2004, 44), because the forestry, fishing, and mining sectors directly supported a range of other businesses in the community – everything from hotels (for an itinerant workforce) to grocery stores (for logging camps) to specialized tradesmen and machinery operators (Young 2006a). Businesses have closed, and people have left the region in significant numbers. Port Hardy's population shrank by nearly a third from 1996 to 2006 (from 5,283 to 3,822). Personal incomes relative to the provincial average have also declined significantly. By 2006, residents were earning an average of $1,100 less than the provincial mean, compared with $3,000 above the provincial mean ten years earlier.

Businesses that remain in Port Hardy face the daunting challenge of re-orienting themselves to deal with profound changes in the local and regional economy. Some have responded by scaling back – reducing expenditures and/or moving to part-time operations – while others have adopted riskier and more aggressive adaptation strategies, such as retraining employees, diversifying what the business does, and trying to attract new clientele inside and outside of the community (see Young 2009).

The aquaculture industry has become a major presence in Port Hardy in the midst of these changes. As we saw in Part 1, this late entry is double-edged. On the one hand, many people credit the aquaculture industry with saving their businesses, and even saving the community from economic ruin. On the other hand, the decline in traditional sectors has meant that many businesses are turning to emerging sectors, such as tourism, that, according to some, are particularly vulnerable to the environmental and

Table 6.12

Types of businesses in Port Hardy, BC, that participated in the study

Sector	Percentage of all businesses
Retail	30.4
Trades	17.1
Tourism	16.0
Hospitality	16.0
In-person services	14.9
Professional services	13.8
Construction	8.9
Utilities	7.2
Other	7.2
Transportation	6.6
Forestry	2.8
Fisheries	2.2
Aquaculture	1.7
Mining	1.7
Recreational services	1.7

Note: Businesses can fall under more than one category.

aesthetic effects of aquaculture (e.g., Marshall 2003). The strength of our research in Port Hardy lies in the fact that we surveyed the entire business community, giving us an opportunity to comprehensively evaluate the impact of the aquaculture industry on different sectors of the local economy.

Table 6.12 shows the demographic details of businesses in Port Hardy that participated in the study. This provides a snapshot of the local economy according to the 65 percent of firms that chose to participate (as with all survey research, we do not know whether unique patterns exist among non-respondents). Most businesses surveyed are involved in the service economy.[15] Notably, few firms indicated direct involvement in forestry, fisheries, aquaculture, or mining, although several such businesses are significant employers in the community. For instance, businesses involved in forestry accounted for a total of 115 jobs in the community, and businesses that indicated direct involvement in aquaculture accounted for 370 jobs.

Our primary measure of the impacts of aquaculture on local businesses involves the statement "I'd like to know if this business has been affected, positively or negatively, by the presence of the aquaculture industry."[16] As shown in Table 6.13, 46 percent of respondents indicated that aquaculture had no effect on their business, the most common answer. Beyond this, however, there is a marked trend towards the positive. Nearly half of all businesses surveyed in Port Hardy (a total of 46 percent) indicated that the

Table 6.13

Effects of aquaculture on local businesses in Port Hardy, BC (self-reported)

The presence of the aquaculture industry has affected this business ...	Percentage
Very positively	17.6
Somewhat positively	28.8
Not at all	46.4
Somewhat negatively	3.3
Very negatively	2.0
Don't know/refused	1.3

presence of the aquaculture industry has had a somewhat positive or very positive impact on them. In contrast, only 5 percent indicated a somewhat negative or very negative impact.

Respondents were also asked an open-ended question – "Can you tell me exactly how aquaculture has affected this business?" – which enables us to investigate how aquaculture is beneficial or harmful to other firms in Port Hardy. Responses were recorded verbatim by interviewers, and were later coded to distinguish between "direct benefits," "indirect benefits," "direct harms," and "indirect harms." We coded a "direct benefit" when a respondent indicated that the business deals directly with the industry in some way. For example, a contract or subcontract with an aquaculture firm is clearly a direct benefit. An "indirect benefit" was coded when a respondent stated something along the lines of, "People are buying more [at my store] because they're working [in aquaculture] and making money" (Respondent 98). The same logic was applied to statements of harm. A direct harm involves any actual loss of business or revenue that a respondent explicitly attributes to aquaculture, whereas an indirect harm involves any statement about general harm (actual or potential) to the economy, wages, or the labour market of the community.

A total of fifty-two businesses indicated that they received a direct economic benefit from the aquaculture industry, and twenty-four others indicated that they benefited indirectly. Only four businesses stated that they had suffered a direct harm, and four others that they had been indirectly harmed.

The number of firms indicating that they benefit directly from the aquaculture industry is surprisingly high. As discussed earlier, there is evidence that large and small aquaculture companies are heavy users of contract arrangements, suggesting a significant spillover into other sectors of the economy. The Port Hardy survey strongly reinforces this. The fifty-two firms benefiting directly from the aquaculture industry represent nearly 30 percent of all businesses surveyed. Most of these direct benefits come from formal

Figure 6.1

Services provided to the aquaculture industry by businesses in Port Hardy, BC

Accounting	Land surveying
Advertising	Legal services
Air transportation	Machine works and repair
Boat moorage	Machinery parts
Boat repair	Marine conservation
Boat sales	Marine navigation equipment
Carpentry	Marine research
Child care	Printing and stationery
Computing/information technology services	Processing
	Road building
Construction	Security
Electronics supply and maintenance	Storage
Fuel supply	Tires
Ground transportation	Waste management
Hardware supply	Welding
Hotel and accommodations	Workplace health and safety

contracting or subcontracting arrangements, or from less formal understandings between local businesses and aquaculture companies about service provision.

Figure 6.1 summarizes the services that Port Hardy businesses provide to the aquaculture industry, as brought to light by the survey. Some are expected, such as boat moorage and fuel supply. Others are less intuitive, such as advertising, child care, computing and information technology, and printing and stationery. Clearly, the aquaculture industry is having a significant impact on a broad range of other businesses in Port Hardy. What is particularly notable about this list is the number of services that could easily be provided outside the community. For instance, accounting, advertising, legal services, and information technology services are not tied to place. Although it is likely that large aquaculture firms also have accountants, lawyers, and computing professionals working outside Port Hardy, the fact that these services are also purchased in an isolated coastal community is significant, suggesting a level of engagement with the region that goes well beyond questions of employment. In Port Hardy, at least, the aquaculture industry is heavily integrated into the local economy. For many businesses, aquaculture has been a lifeline:

> With everything else in the doldrums – [aquaculture] has saved our behinds, I can tell you that much ... We'd be tanked if we didn't have them. (Manager, Professional services firm: Interview reference 136)

> The economy around here has been so unstable – aquaculture is keeping it somewhat stable. It's something that gives the economy so much support. This has been a much needed and appreciated investment [in Port Hardy]. (Manager, Trades firm: Interview reference 342)

A small percentage of business operators are not enthusiastic about the aquaculture industry, however. Most of the roughly 5 percent of respondents that experienced a direct or indirect negative impact from aquaculture are related to tourism. To be clear, this is a still a minority opinion: among tourism-related businesses in Port Hardy, 56 percent indicated that they were "not at all" affected by the aquaculture industry, 22 percent indicated a positive effect, and 22 percent a negative effect. Those that indicated a negative effect tended to be involved in expedition tourism or the sport fishery. Their criticisms have a distinct environmental theme:

> Aquaculture is potentially very negative. We're concerned about the sea lice issue. It will devastate wild salmon stocks. And if that happens, it will directly affect [our business]. (Owner, Tourism firm: Interview reference 240)

> We have a lot of recreational fishermen [as customers]. If there's lice on the fish, or if they're catching [escaped] Atlantic salmon, then that's not good for them or for us. (Manager, Tourism firm: Interview reference 7)

Tourism operators are not alone in criticizing the aquaculture industry. Several businesses that formerly serviced or supplied the commercial fishery also indicated a negative impact. Again, however, these voices were relatively rare:

> There is less attention paid to commercial fishing because of aquaculture ... We do get some [aquaculture] boats coming in for [supplies], but these boats used to be fishing boats. I feel that the federal and provincial [government] policies support fish farmers. They are throwing money at the fish farming industry to get them going and to take over markets held by salmon fishermen. (Owner, Retail firm: Interview reference 253)

Overall, it is fair to conclude that the business community in Port Hardy is very positive towards the aquaculture industry. A surprising (to us) number of businesses indicate that they benefit directly from the industry – mostly by providing goods and services directly to aquaculture firms in the region. Negative impacts and opinions were much rarer than we expected, and were expressed in the expected sectors, particularly tourism.

Finally, though, it is important to note that some respondents were lukewarm about the aquaculture industry, even though their businesses were clear beneficiaries. For example, one respondent, whose firm holds a contract with an aquaculture company, stated:

> Yes, we do do work for the fish farms. That industry is going to create work whether we like it or not. They're a business [like any other] and they need [our service]. So we provide it. (Owner, Trades firm: Interview reference 263)

The manager of another firm, also with strong connections to the industry, is even more ambivalent:

> [I think] there has been a lot of negative impact [from aquaculture], but not for our business. Aquaculture has taken the place of traditional fisheries as far as this business is concerned ... They say fish farming is on the way out because of concerns for the environment. That would be bad for business but I would say, environmentally, that would be a good thing. (Manager, Trades firm: Interview reference 387)

Summary: Aquaculture and Coastal Development

We began this chapter by highlighting a dilemma facing federal and provincial governments in Canada regarding the aquaculture industry. On the one hand, aquaculture has great potential in this country. The industry is anxious to expand, and Canada's vast coastlines are well suited to aquaculture production. On the other hand, there is recognition that expansion must proceed "in harmony with the social fabric of [coastal] communities" (Bastien 2004, 3). Coastal regions in Canada have endured significant economic restructuring and decline over the past few decades. For many industry supporters, aquaculture offers a way to reverse the human and economic tide flowing away from coastal regions, but it also remains deeply controversial at the local level (see Chapter 2).

We have attempted to measure the impact of aquaculture on community economies. Aquaculture, like all industries, is disruptive. It affects local labour markets, makes claims on local spaces and environment, and in some cases competes with traditional commercial fisheries in national and global markets. In the process, it creates winners and losers. Industry supporters argue that the benefits outweigh the costs: specifically, that the industry provides much-needed employment and has spillover effects on other businesses in depressed coastal regions (Bastien 2004; NBSGA n.d.a). Critics, on the other hand, argue that the industry harms or at the very least threatens other sectors of the economy (Marshall 2003; Cox 2004).

Our research has revealed a mixed landscape, with some significant findings regarding the positive impacts of aquaculture development. Regarding employment, we conclude that some estimates of the impact of aquaculture are highly optimistic. Our nationwide survey of aquaculture firms showed that direct employment in aquaculture in 2003 was likely 5,000 to 6,000 full-time, part-time, and seasonal or casual jobs – significantly lower than the estimates of 12,000 jobs that appear in some industry publications (e.g., CAIA 2006a). The survey also indicated, however, that most jobs in the industry are full-time and year-round. This stability is notable in coastal regions where many other employment sectors are highly seasonal.

Other important trends have emerged from this research. Aquaculture is a big employer of youth and young adults but not older workers; of men but not women; of non-Aboriginal but not Aboriginal people. There is also a clear divide between salmon and shellfish aquaculture. Despite often-heard claims about small businesses driving employment, the survey shows that large companies are the big employers, and these are overwhelmingly salmon aquaculture firms. They also tend to pay higher wages, provide better employee benefits, and engender higher levels of job satisfaction in workers.

Overall, we conclude that there are two faces to aquaculture employment. On the one hand are "core jobs" – stable full-time jobs with modest to high levels of compensation and a satisfied workforce. On the other hand are "contingent jobs," which are more unstable, generally poorly compensated, and less fulfilling. It is particularly distressing to note that Aboriginal people are over-represented in the contingent workforce and under-represented in the core workforce (see Table 6.7).

We also investigated the relationship between aquaculture firms and communities. There is some evidence of tension here. In particular, many firms expressed frustration with the work ethic and commitment of the local labour force. At the same time, there is evidence of a strong positive impact of aquaculture on community economies. Our case study of Port Hardy, British Columbia, showed that the industry has a broad impact on the local economy and that its presence directly benefits a surprising number and range of businesses (see Figure 6.1). This suggests that aquaculture is a force for economic diversification. As we mentioned at the outset of this chapter, diversification is widely accepted as the only means of ensuring community stability and prosperity in coastal communities in "post-resource" economies. If aquaculture firms have direct links to local tradespeople as well as accountants, construction firms as well as child care centres, then this industry makes a considerable contribution to both stability and diversification in local economies.

Finally, our research in Port Hardy found significant support for the aquaculture industry across the business community. Nearly half of all businesses indicated that they benefited from the presence of the aquaculture industry, compared with only 5 percent indicating a negative impact. Opinion among tourism operators was evenly split between positive and negative impacts, despite the fact that critics argue that tourism is particularly vulnerable to the environmental and aesthetic impacts of aquaculture.

The Port Hardy research is a case study, which means that it is difficult to generalize our findings. Undoubtedly, critics of aquaculture would point out that this research was conducted in a place that enjoys the benefits of employment and contracts from the industry, while the costs (particularly to the environment) are being felt far outside the community (see Marshall 2003, 22). Aquaculture supporters would likely argue, however, as did one respondent to the Port Hardy survey, that "there's a lot of untapped potential out there for aquaculture to do a lot of good for [other] communities if they were just allowed to expand there too."

7
Governing Aquaculture

Federal and provincial governments in Canada are deeply committed to aquaculture development. Like a mirage in the desert, aquaculture tantalizes state authorities as a potential solution to some serious political and economic problems: a way to spur growth in isolated areas that are suffering significant economic hardship; a way to shift the coastal labour force away from declining sectors and environmentally damaging activities; a means of re-establishing Canada's presence in the growing global seafood market; a way of escaping Canada's "staples trap" dilemma by establishing a new high-value and high-productivity industry driven by research and development; and, yes, the promise of achieving all this in an environmentally sustainable manner. These as yet unrealized promises have spurred Canadian governments to throw their weight behind aquaculture in a big way.

Aquaculture has also been a policy nightmare, however. Governments are navigating uncharted waters in their attempts to simultaneously promote, regulate, and legitimize the aquaculture industry (Rayner and Howlett 2008). As discussed in Part 1, aquaculture is the first major resource sector in Canada to grow up in the new realities of economic globalization, heightened public and stakeholder sensitivity to environmental degradation, and powerful Aboriginal and stakeholder rights movements. Adding to the challenge is that aquaculture takes place in a jurisdictional grey area. The *Constitution Act, 1867*[1] states that oceans and inland fisheries are the responsibility of the federal government, as are issues of marine navigation, shipping, and Aboriginal affairs, while the provinces are responsible for environment, the management of Crown lands, private property and business regulation, and rivers and tidal areas in bays, inlets, and estuaries (cf. Howlett and Rayner 2004). While federal and provincial governments have worked to disentangle these jurisdictional overlaps through memoranda of understanding, joint federal/provincial committees, and other agreements, aquaculture law and regulation is still much more complicated in Canada than in other nations unburdened by this tradition of federalism and unique separation of powers (Tollefson and Scott 2006). Finally, aquaculture policy in Canada is complicated by the ongoing public controversy over the industry. Both the aquaculture industry and state authorities have clearly been taken aback by the persistent and vocal criticism from opponents of aquaculture development (see Chapter 3). While governments have rarely responded directly to these criticisms, opposition has prompted federal and provincial governments to make "building public confidence in the industry" a policy priority (DFO 2000).

In this chapter, we examine how federal and provincial governments in Canada are seeking to govern aquaculture under these complex pressures. We argue that governments are using a blend of traditional and non-traditional policies in their efforts to promote, regulate, and legitimize the industry. The most significant traditional policy response to aquaculture has been to aggressively subsidize firms and industry associations. Millions of government dollars have been spent on direct industry support programs, as well as on indirect support for research and development. Canada has a long history of subsidizing resource and primary industries, and governments across the country have returned to this well with enthusiasm. As we shall see, these subsidies are intended to improve the competitiveness of Canadian industry by reducing lending and other operating costs and by encouraging technological innovation and investment. Subsidies, however, do little to address other important problems facing the industry, namely, environmental performance (real and perceived) and the legitimacy of the industry in the eyes of stakeholders and the general public. We therefore argue that governments are also applying non-traditional policies to aquaculture that are intended to address questions of competitiveness, environment, and legitimacy in novel ways. These newer strategies vary by jurisdiction and region, but share a preoccupation with broadening the governance tent to include (and implicate) nongovernmental actors – particularly firms, industry associations, and stakeholders – in aquaculture policy and regulations. These strategies have various labels, including "governing in partnership," "new public management," "devolution," "smart regulation," "results-based regulation," and "performance-based regulation" (see Gunningham et al. 1998; Doern and Reed 2000; Vold 2003; Howlett and Rayner 2004). These newer policies are controversial. Some critics see them as a step towards deregulation, while advocates argue that they are a way to improve the industry's bottom line while improving environmental compliance and enhancing its legitimacy.

Problems of Aquaculture Governance

Governing aquaculture in Canada is not a simple task. As mentioned, the jurisdictional overlaps between federal and provincial responsibilities pose a significant challenge for both levels of government (Tollefson and Scott

2006). Despite efforts to harmonize policies and regulations and enhance interagency communication, there remains significant overlap. For instance, although environment is a provincial responsibility, many aquaculture operations are subject to the *Canadian Environmental Assessment Act*,[2] and the federal *Fisheries Act*[3] requires the Department of Fisheries and Oceans (DFO) to engage in its own review of whether an aquaculture site or operation would "cause a harmful alteration, disruption, or destruction of fish habitat" (Section 35[1]).

Besides jurisdiction issues, aquaculture policy and regulation are still works in progress. The rules governing aquaculture in Canada have rarely stood still, and have been overhauled numerous times at both provincial and federal levels. Following the emergence of large-scale commercial aquaculture in the 1980s, many provinces scrambled to introduce legislation to govern the new industry. Nova Scotia, New Brunswick, and Newfoundland all passed *Aquaculture Acts* during this early period, although both Nova Scotia and New Brunswick later substantially reformed these laws (in 1996 and 2000, respectively) in the interests of streamlining regulations and promoting industry expansion (Phyne 1996; Marshall 2001; Howlett and Rayner 2004). In British Columbia, aquaculture regulations fall under multiple ministries. At the time of writing (2008), the Ministry of Agriculture and Lands (MAL) and the Ministry of Environment (MoE) have primary responsibility. Regulations in the province have been in constant flux since the early 1980s, however. The industry has now been subject to three policy reviews (the Gillespie report in 1986, the Salmon Aquaculture Review in 1996-97; and the Special Committee on Sustainable Aquaculture in 2007-08) – each of which has made major recommendations for policy and regulatory change. As we shall discuss later, British Columbia has been one of the most aggressive innovators in aquaculture policy, particularly since the 2001 election of the BC Liberal Party (BCLP) under Premier Gordon Campbell. At the federal level, the government of Canada regulates the aquaculture industry primarily under the *Fisheries Act,* but over the years has released a dizzying array of policy statements and strategies for aquaculture development. As we shall also discuss, the federal government has legal obligations to play a strong role in aquaculture regulation, but it has also been active in other areas, such as directly subsidizing the industry, working to reduce jurisdictional overlap, and sponsoring research and development.

Despite the best efforts of the two levels of government, few stakeholders are pleased with how aquaculture is regulated. Both supporters and opponents of the industry regularly complain about aquaculture policy, albeit for very different reasons. For instance, the aquaculture industry regularly claims that it is excessively burdened by regulations and requirements. According to one study conducted in 2004, "the British Columbia aquaculture industry is subject to 52 separate federal and provincial statutes, regulations, policies and guidelines, as well as numerous municipal and regional land use and development regulatory processes" (Gislason and Associates 2006, 12). The British Columbia Salmon Farmers Association (BCSFA) reports that its members regularly invest several years and up to $200,000 in preparing multi-jurisdictional applications for a new site (Robson 2006, 26). Critics argue, however, that the aquaculture industry enjoys an overly cozy relationship with government (Cox 2004; Langer 2004; Peterson et al. 2005). They are particularly upset by government funding for industry associations such as the BCSFA and the Canadian Aquaculture Industry Alliance (CAIA). They are also concerned with the involvement of the Department of Fisheries and Oceans in aquaculture development. DFO's traditional mandate has been the management and conservation of wild fisheries and navigable waters, but since 1984 it has been the lead federal agency for aquaculture regulation and development, a role that some see as fundamentally contradicting its conservation mandate (e.g., Langer 2004; Morton 2004). The close relationship between industry and government has prompted accusations that regulators bury unfavourable research and monitoring data, avoid fining violators for malfeasance, and permit aquaculture firms to circumvent key environmental regulations (e.g., Langer 2004, 131; Cox 2004, 65-66).

There is also significant disagreement about which laws and government programs should apply to aquaculture. Although environmentalist groups are critical of DFO, many believe that aquaculture should be regulated under the *Fisheries Act* and other regulations aimed at conserving wild fish populations. For example, the David Suzuki Foundation has repeatedly argued that DFO needs to base aquaculture policy on the *Fisheries Act* instead of developing new policies and regulations specifically for the industry (Tank 2004; Young and Werring 2006). Conversely, some aquaculture advocates argue that the industry should be regulated as though it were equivalent to land-based agriculture (Bastien 2004; Halse n.d.). Yves Bastien, the former federal Commissioner for Aquaculture Development, has strongly supported this idea. His report *Recommendations for Change* stated: "Aquaculture has been somewhat of an orphan within [DFO], under-resourced and never fully understood. Indeed, aquaculture has been at the centre of an internal conflict within DFO for some time. Perhaps we must ask whether the federal government is taking the correct approach with respect to this industry ... We continue to treat [aquaculture] as a subset of the fishery rather than as the agri-food business that it is" (Bastien 2004, 31). Aquaculture advocates argue that land-based farmers enjoy significant government supports that are not available to aquaculturists, including state-backed insurance in the instance of crop failures or disease outbreaks, exemption from certain municipal nuisance laws (so-called right-to-farm legislation that exists in many Canadian provinces), and less stringent waste management regulations (Bastien 2004; Halse n.d.).

Finally, the challenges of governing aquaculture are exacerbated by three overarching problems that we call the "triple pressure." The first problem is economic. There is an acute awareness in government circles that economic globalization is putting tremendous competitive pressure on the aquaculture industry in Canada. International commodities markets, including the seafood market, have become very competitive, and Canada's traditional reliance on high-volume/low-value production and export to the US market is an increasingly risky strategy (cf. Burda and Gale 1998; Hayter 2000; Knapp et al. 2007). The second problem is environmental. As discussed in Chapter 1, the environment has become a major political issue in Canada. Although this pressure is exerted unevenly (as we just saw, aquaculturists look enviously at farmers, who routinely alter landscapes and environments without protest), environmental recklessness is increasingly unacceptable to stakeholders and the general public. The third problem is one of legitimacy. Governments are having a difficult time legitimizing the aquaculture industry, or their own policies and regulations towards it.

In the rest of this chapter, we consider how governments in Canada are responding to competing demands regarding aquaculture policy, as well as to the triple pressure of economy, environment, and legitimacy. We will argue that, in general, governments are responding in very traditional ways (creating generous and extensive subsidy programs) as well as in distinctly non-traditional ways that seek to address problems in aquaculture governance by "going beyond government."

Traditional Policy Responses: Subsidy and Sponsorship

Canadian governments have long subsidized resource development. These subsidies can be direct, as when governments provide funding to firms and producers, or indirect, as when governments forgive or minimize resource rents or taxes (Ragosta 1990). The provision of subsidies is controversial. Some critics argue that they are distortions on the market, and as such they have been targeted by free trade advocates around the world (Pearce and von Fickstein 2002). Others argue that subsidies have played a role in the overexploitation of resources by encouraging investment in industries that should otherwise be permitted to decline (Myers and Kent 2001). In Canada, however, subsidies have long been seen as an important tool for regional and economic development (Matthews 1983). Two main justifications are given. On the one hand, many of Canada's resource industries are very capital-intensive (particularly forestry, mining, and oil and gas), and Canadian governments have justified subsidies to these sectors as a means of either attracting foreign investment or encouraging the development of Canadian firms (cf. Safarian 1973). On the other hand, subsidies have been directed to industries such as commercial fishing and agriculture primarily for political reasons, chiefly to sustain employment in rural and isolated regions of the country (Grafton and Lane 1998).

Table 7.1 outlines the striking array of direct and indirect subsidies to the aquaculture industry from Canadian governments. Both federal and provincial governments are active in subsidizing research and development, providing grants and loans to individual firms, and assisting in the acquisition of skills and technologies. As discussed in Chapter 1, the federal government in particular is committed to a strong research and development strategy for aquaculture in the hopes of emulating the Norwegian "high road" to aquaculture development, where wealth is generated through innovations, value-added production, patents, and "knowledge exports" rather than low-cost production (Knapp et al. 2007). Nevertheless, governments are also clearly committed to traditional firm-targeted subsidies that have long characterized Canadian resource development.

Non-Traditional Policy Responses
As discussed earlier, governments in Canada face a difficult "triple pressure" with respect to aquaculture: ensuring that the Canadian industry is globally competitive, minimizing harm to fragile and damaged ecosystems, and building legitimacy for the industry in the face of local and organized opposition. The litany of subsidies to the aquaculture industry is clearly aimed at enhancing industry competitiveness by reducing costs, sponsoring innovations, and encouraging risk taking. Subsidy programs do little, however, for the crises of environment and legitimacy facing the aquaculture industry in Canada (although some subsidies are directed towards research into environmental impacts and the development of new environmental technologies).

In this section, we examine some key policy innovations that are being applied to aquaculture governance in this country. We use the term "policy innovations" generously, because many of the trends that we identify are not unique to aquaculture and are evident in other sectors and jurisdictions in the world. These policy changes are generally understood to break from what has been called a command-and-control approach to environmental and resource management. The command-and-control approach is characterized by strong state involvement in governing the environment and resource development, typically by establishing "a complex web of legislation, agency rules, permit procedures, standards, judicial decisions, and other enforceable environmental policies, underpinned by a variety of sanctions" (Sinclair 1997, 529). In other words, a command-and-control approach is based on heavy government involvement in rule setting and rule enforcement.

Recently, there has been a global trend away from the command-and-control approach, particularly in Western democracies in Europe, North America, and the South Pacific. Despite disagreement over what exactly is

Table 7.1

Direct and indirect subsidies to the aquaculture industry from Canadian governments

Agency	Purpose	Amount of subsidy
Federal government		
Atlantic Canada Opportunities Agency	"The development of salmon aquaculture on the South Coast of Newfoundland"	$9.4 million
Western Economic Diversification	"The promotion of technology development and transfer to the aquaculture industry"	Varies; $250,000 to Kitasoo/Xai'xais First Nation
Farm Credit Canada	Loans to individual firms	$30 million per annum
Aquaculture Partnership Program (OCAD)	"To assist the development of the Canadian aquaculture industry"	$600,000 per annum
Indian and Northern Affairs Canada	Grants and loans to Aboriginal aquaculture businesses	Varies
Aquaculture Collaborative Research and Development Program (DFO)	"To increase the level of collaborative research and development activity between the aquaculture industry and DFO"	$4.5 million per annum
AquaNet (DFO, NSERC, SSHRC)	"To foster a sustainable aquaculture sector in Canada through high quality research and education"	$14.4 million over 4 years (expired)
Newfoundland and Labrador		
Aquaculture Working Capital Loan Guarantee Program	Loans	Varies
Aquaculture Capital Incentive Program	"To increase production from both hatcheries and marine sites"	Varies; up to $250,000 for finfish and $100,000 for shellfish firms

Program	Description	Amount
Aquaculture Innovation Program	"To assist commercial aquaculture operations to evaluate, develop, and adopt new technologies"	Share of $12.5 million per annum
Human Resources Development Program	"To provide financial support for companies to upgrade the knowledge and skills base of employees"	Share of $12.5 million per annum
Marketing Intelligence Program	"For the investigation of specific areas of opportunity for aquaculture products"	Share of $12.5 million per annum
Nova Scotia		
Fisheries and Aquaculture Loan Board	Loans	$25 million per annum
Fisheries and Aquaculture Development Program	"For new or improved harvesting and processing technology, aquaculture development, and new community-based infrastructure"	Up to $50,000 per firm
Prince Edward Island		
Shellfish Aquaculture Financing Program	"Providing funding to shellfish growers in order to facilitate expansion of production"	Varies
Aquaculture Technology Program	Grants for equipment upgrades	Up to $10,000 per firm
Aquaculture Diversification Program	Grants "for the creation of a new enterprise or the diversification of an existing one"	Up to $10,000 per firm
Aquaculture and Fisheries Research Initiative	Grants "for research projects initiated by industry associations, private businesses, educational institutions, and government agencies"	Up to $60,000 per project
New Brunswick		
Loan Guarantee Fund	Loans "to aid and encourage the development of aquaculture"	Up to $500,000 per firm
Total Development Fund	"To provide 'top-up' funding to major aquaculture projects'"	Varies

▼ *Table 7.1*

Agency	Purpose	Amount of subsidy
Quebec		
Mariculture Industry Development Corporation	Grants, loans	Varies
Continental Aquaculture Research and Development Corporation	Research and development assistance	Varies
Innovation and Development Support Program	"For marketing development, biotech research, and innovations in processing and marine technology"	Varies
British Columbia		
Shellfish Aquaculture Working Capital Fund	Loans	Varies
Aquaculture and Environment Fund	"For research into the environmental, economic and social performance of the industry"	$3.75 million
Malaspina University-College	Training for Aboriginal people in shellfish aquaculture	$2 million
University of British Columbia, Chair in Sustainable Aquaculture	Research	$1.25 million
Various agencies	Research	$5.1 million

Sources: WED 2000, 2007; OCAD 2002a, 2002b; British Columbia 2002a, 2002b; INAC 2003, 2005; Farm Credit Canada 2004; Howlett and Rayner 2004; AquaNet 2005; Environmental Law Centre 2006; Newfoundland and Labrador 2007; ACOA 2008; DFO 2008; Networks of Centres of Excellence Program 2008.

replacing command and control in these cases, many observers argue that governments are increasingly looking for ways to "go beyond government" and attempt to involve non-state actors in the formulation and enforcement of policy (e.g., Rose 1999; Dean 1999; Peck and Tickell 2002; Howlett 2003; Levi-Faur 2005). With respect to aquaculture policy in Canada, we argue that this is being done in two ways. First, governments are engaging in legitimacy-building activities that include important new citizen engagement initiatives. Second, governments are devolving significant authority and responsibilities to aquaculture firms and industry associations. This is being done in an attempt to give aquaculture companies greater flexibility in meeting environmental standards (thereby improving their bottom lines and, according to advocates of this approach, improving environmental compliance at the same time). Both strategies are in their infancy and are being haltingly and unevenly applied across jurisdictions, but we will argue that both are crucial in addressing the triple pressure that threatens to sink the aquaculture industry.

Legitimacy-Building Activities
In the early years of the industry, Canadian governments assumed that aquaculture would take its place alongside other coastal industries as a legitimate user of ocean space and resources (cf. Keller and Leslie 1996; Marshall 2001). Recently, however, they have become acutely aware of the severity of the industry's legitimacy problem. In 2001, the Canadian Council of Fisheries and Aquaculture Ministers – a body composed of all federal and provincial ministers responsible for fisheries and aquaculture – released a "Canadian Action Plan for Aquaculture" that identified public skepticism of aquaculture as the most serious barrier to industry development (CCFAM 2001). A year earlier, DFO had released an "Aquaculture Action Plan" built around two policy goals: "increasing public confidence in the sustainability of aquaculture development" and "increasing industry competitiveness in global markets" (DFO 2000).

Aquaculture's legitimacy problem will not be easy to fix. As we have seen throughout this book, the industry's failure to gain broad public acceptance is complicated, having to do with coincident declines in traditional rural livelihoods, competing scientific conclusions and narratives, media and activist discourses, and the inability of aquaculture supporters to assuage public and stakeholder fears. Governments are also contributing to the legitimacy crisis that they are trying to alleviate. Opponents of aquaculture are particularly galled by the myriad of subsidies that have been showered on the industry by governments eager to promote development (e.g., Marshall 2003; Cox 2004).

Howlett and Rayner (2004) argue that, in a post–command-and-control era, governments are increasingly relying on "procedural instruments" in order to build legitimacy for their decisions, and by extension for the industries that they are regulating. Procedural instruments are any formalized processes of decision making or information gathering in formulating or implementing policy. Under traditional command-and-control regimes, procedural instruments were typically internal to government. Lawmakers would consult with experts housed in government departments and negotiate policy among themselves. According to Leiss (2000, 49-50):

> In the older [command-and-control] model, governments thought it wise to provide for in-house science expertise so as to translate risk assessments into policy choices ... Like everything else in the civil society bureaucracy, the scientific advice was given in secret and was thus not open to challenge by the public, which was especially useful for covering up the existence of huge uncertainties and the lack of essential data to support a preferred policy choice.

While this method is still standard in many branches of government, it is no longer viable in highly controversial sectors such as aquaculture (see Chapter 3). According to many scholars, governments are increasingly turning to new procedural instruments that *include outside actors* in order to legitimize their decisions (Irwin 1995; Gunningham et al. 1998; Howlett 2000; Leiss 2001). Such procedural instruments can include holding public and/or expert hearings, engaging in one-time or ongoing consultations, establishing multi-interest groups and committees, creating arm's-length advisory committees, and directly sponsoring interest and stakeholder groups (Howlett 2000).

Howlett and Rayner (2004) argue that, despite the severity of aquaculture's legitimacy crisis, Canadian governments have been slow to develop new procedural instruments in this sector (see also Rayner and Howlett 2008). Although the federal government has been very active in subsidizing aquaculture development, its procedural activities have been limited. Its most significant activity to date has been to provide funding for the development of industry associations such as CAIA to become liaisons (and lobbyists) between government and industry. This has been useful in building legitimacy among members of the aquaculture industry, but it has not sat well with other stakeholders (Howlett and Rayner 2004).[4]

While federal procedures remain underdeveloped, several provinces have implemented formal processes as means of including outside actors and building legitimacy. For instance, some jurisdictions have experimented with community-level involvement in aquaculture regulation. The most significant of these is in Nova Scotia, which has implemented Regional Aquaculture Development Advisory Committees (RADACs) in twelve coastal communities and regions. RADACs have limited powers (essentially to review and comment on development proposals) and limited mandates: "to facilitate economic development while simultaneously providing information to local residents and determining the level of support [for a given project]" (Nova Scotia 2007). RADAC members are "appointed by the Minister, and [committees] are composed of people who represent the interests of the area. This may include fishermen, aquaculturalists, recreational boaters, waterfront landowners, business operators, local politicians, and others – in short, people and groups affected by the installation of an aquaculture site" (Nova Scotia 2007). Although RADACs do not have the power to stop or modify aquaculture projects, their conclusions are passed on to the provincial minister as recommendations.

Similarly, Prince Edward Island has established a Shellfish Advisory Committee (SAC), which mirrors the RADAC program but at a smaller scale. The SAC is composed of commercial fishers, federal and provincial regulators, representatives of Aboriginal communities, and aquaculture processors (Howlett and Rayner 2004). It is chaired by a representative of DFO, and meets at least once per year or "as issues come up which need industry feedback" (DFO 2006e).

In British Columbia, several efforts have been made to involve communities and stakeholders in aquaculture development and regulation. The provincial government has been particularly anxious to involve communities and First Nations groups in shellfish aquaculture. In 1998, it announced a Shellfish Development Initiative (SDI), with the stated aim of "doubling the amount of Crown land available for shellfish aquaculture" (British Columbia 2007a; Rayner and Howlett 2008). To achieve this, the province committed to "working with coastal and First Nations communities and industry to determine acceptable siting and production levels for shellfish aquaculture within each community." The SDI currently sponsors fifteen pilot projects "in cooperation with local First Nations" in northern coastal regions of the province (British Columbia 2007a). In addition, in 2005 the province offered to sign memoranda of understandings (MOU) with First Nations groups that were interested in pursuing shellfish aquaculture tenures. The MOUs give exclusive rights to First Nations to apply for tenures and licences in their traditional territories for a period of ten years (Tollefson and Scott 2006, 33). In 2007, the province moved to reduce application fees for shellfish tenures in order "to attract new producers, including First Nations" (British Columbia 2007b).

While procedural instruments are essential for building legitimacy, they can also be unpredictable and double-edged. Governments take risks when they establish means for communities, stakeholders, activists, experts, and politicians to speak out on or participate in aquaculture policy. For example, the government of British Columbia has engaged in two high-profile public and expert reviews of aquaculture, and both have had mixed outcomes as legitimacy-building exercises. The first, the Salmon Aquaculture Review (SAR), was directed at salmon aquaculture alone.[5] It was initiated in 1995 under the left-of-centre New Democratic Party (NDP) government with a mandate to "conduct a review of the adequacy of current methods and processes used in regulating and managing aquaculture operations in British Columbia" (SAR 1997a, 2). The SAR was led by the province's Environmental Assessment Office, a branch of the government that "operates independently and coordinates the assessment of major projects in British Columbia." It undertook reviews of aquaculture technology and socioeconomic issues, and also received submissions from First Nations groups and other stakeholders. It resulted in forty-nine recommendations on improving aquaculture policy and regulations. Many industry supporters saw this as a green light to proceed with expansion (Gardner and Peterson 2003, 1), but opponents have based much of their subsequent activism on the SAR, accusing the province of failing to address the concerns raised in that process (e.g., Georgia Strait Alliance 2004).

Ten years following the SAR, the right-of-centre British Columbia Liberal Party struck the Special Committee on Sustainable Aquaculture (SCSA). The SCSA was composed of ten Members of the Legislative Assembly and held public hearings in nineteen communities and cities on the Pacific coast. It was unusual in that it was chaired by an opposition NDP member, and BCLP members were outnumbered on the committee. Ultimately, the SCSA made two controversial recommendations: that salmon aquaculture move to "ocean-based closed containment systems," where ocean net-based pens are replaced by a solid barrier, and that the provincial government not permit the salmon aquaculture industry to expand into northern coastal regions. To date, the provincial government has not responded to the first recommendation. After a lengthy delay, however, it announced a moratorium on northern expansion of salmon aquaculture "to allow time to explore new management options for aquaculture policies" (British Columbia 2008b).

Ongoing efforts to engage communities and stakeholders are also double-edged. In British Columbia, the provincial government announced in 2001 that it would establish Community Shellfish Steering Committees that would be similar to Nova Scotia's RADACs. According to the government, however, the committees often proved obstructionist and "impossible to work with," and were soon disbanded (Howlett and Rayner 2004). The province also established a Salmon Aquaculture Implementation Advisory Committee (SAIAC) in 1999 "to involve First Nations, coastal communities, environmental organizations, industry and the federal and provincial governments in the implementation of regulations, policy development, and the strategic development of the salmon farming industry" (British Columbia 2002d, 6). Members included many prominent activists (environmentalist, Aboriginal, and industry) as well as municipal leaders and commercial and sport fishers. The committee fell apart in 2002 following the province's decision to lift the moratorium on new salmon farming tenures, which prompted several First Nations members and representatives of environmentalist groups to resign. According to some former members, the SAIAC "was left out of the loop on every major decision made by the province on salmon aquaculture," and "all key decisions on salmon farming were made behind our backs with the salmon farming industry" (Georgia Strait Alliance 2002)

Devolution to Industry

A second series of policy innovations has to do with devolution to the aquaculture industry. Devolution is a complex policy tool. Generally speaking, it refers to the transfer of authority and/or responsibilities between governments (usually from federal to provincial governments, or from provincial to local governments), or from government to nongovernmental actors such as firms, groups, and individuals (Young and Matthews 2007b). As such, devolution is often associated with deregulation or with the offloading of responsibilities from governments that want to wash their hands of particular obligations. Devolution can also be a way of expanding or reforming regulations, however, and many policy observers argue that it is becoming an increasingly common strategy in environmental and resource management as an alternative to the traditional command-and-control approach that we discussed earlier (e.g., Walker and Crowley 1999; McCarthy 2006).

Political scientists and policy makers have applied several terms to devolution-based policy strategies, including "self-regulation" (Sinclair 1997), "results-based regulation" (Vold 2003), and "smart regulation" (Gunningham et al. 1998). There is significant overlap among these terms, as each implies the transfer of authority and responsibilities for environmental integrity away from the state. Generally speaking, "self-regulation" refers to industries' and firms' voluntarily compliance with environmental codes and practices that are often negotiated with government (Sinclair 1997; Harrison 2001):

Whereas previous environmental measures tended to be prescriptive in character with an emphasis on the "thou shalt not" approach, [self-regulation] leans more towards a "let's work together" approach. This reflects the growing realization in industry and in the business world that not only is industry a significant part of the (environmental) problem but it must also be part of the solution. The new approach implies, in particular, a reinforcement of the dialogue with industry and the encouragement, in appropriate circumstances, of voluntary agreements and other forms of self-regulation. (Harrison 2001, 207)

In practice, self-regulation has rarely involved the complete withdrawal of the state (Harrison 2001). Instead, it can involve a range of measures that include "voluntary and cooperative agreements, environmental covenants, co-regulation, enforced self-regulation, negotiated compliance, codes of practice, environmental partnerships, environmental management systems, corporate environmental reporting, environmental accounting, and environmental self-auditing" (Sinclair 1997, 532).

The term "results-based regulation," which also goes by the label "performance-based regulation," is similar but more precise. Advocates of results-based regulation argue that the command-and-control approach puts too much emphasis on regulation of industrial processes. Traditional "process-based regulation" sets specific standards and requirements for how firms conduct themselves. For instance, forestry regulations in Canada have traditionally directed or restricted firms in their use of equipment, methods for site preparation and clean-up, roadbuilding, mix of species cut, and schedules for reforestation (Martin et al. 2002; West Coast Environmental Law 2002). A results-based regulatory system focuses instead on setting standards for *final outcomes*, while granting firms greater discretion in deciding how to achieve them (cf. Durant et al. 2004). "Under a results-based approach, government clearly defines outcomes and companies decide how to achieve them. This gives firms and the resource professionals they hire more freedom to manage, with government holding them accountable for achieving the defined results" (Vold 2003, 1).

"Smart regulation" is a popular term in government circles and among political scientists. It is used to describe "new [methods] of governance that ... promote looser regulatory arrangements in which governments attempt to steer networks of actors in a desired direction rather than manage the sector through traditional policy instruments" (Howlett and Rayner 2004, 172). Smart regulation is about "steering" and "encouraging" certain actions among industry and other stakeholders rather than prescriptively governing them. In this respect, smart regulation has much in common with results-based regulation. Indeed, in 2004, the government of Canada struck an External Advisory Committee on Smart Regulation (EACSR), which defined smart regulation in a way that strongly mirrors results-based regulation: "Smart regulation is more responsive regulation. An effective regulatory system must be self-renewing and keep up with developments in science, technology, and global markets ... This means giving regulators more flexibility in terms of how results are achieved, as long as high standards are upheld and the appropriate accountability measures are in place" (EACSR 2004, 12).

Advocates of smart regulation argue that governments need to give up significant authority and engage nongovernmental actors as "partners." Gunningham and colleagues (1998, 10) state:

Government interventions [need to be more] selective and combined with a range of market and non-market solutions, and of public and private orderings. Given the political and fiscal constraints under which governments now labour, it will be crucial to harness resources outside the public sector in furtherance of government policy ... Among these opportunities are those than flow from market phenomena, and from commercial influences. These and other mechanisms may be enlisted in furtherance of "governing at a distance."

The main idea here is that governments need to shift certain functions and responsibilities to nongovernmental actors in order to enhance the efficiency, flexibility, and legitimacy of regulatory regimes. The key is cooperation between government and the governed. According to the EACSR (2004, 9):

Canada's regulatory system [must be] improved in order to sustain Canada's well-being in the future. This objective cannot be realized without cooperation among governments, industry and citizens, which is why cooperation is at the heart of the Committee's proposals ... The Committee believes strongly that a high-performing 21st century regulatory system requires better and closer relations between the partners in the system – governments, government departments, industry, citizens/consumers and other stakeholders – based on improving information, transparency and trust.

Governments in Canada have recently begun to apply policies in the aquaculture sector that reflect the principles of self-regulation, results-based regulation, and smart regulation. As discussed earlier, governments have a constitutional obligation to strongly regulate aquaculture because it takes place in public waters. Nevertheless, in recent years both federal and provincial governments in Canada have made commitments to work more closely with industry, to give firms greater discretion over environmental compliance, and to make aquaculture policy more flexible and responsive to industry concerns. We shall discuss federal and provincial efforts to move in these directions in turn.

As mentioned earlier, DFO has been the lead federal agency for regulating aquaculture in Canada since 1984. It regulatory approach is governed by three key federal statues: the *Fisheries Act,* the *Navigable Waters Protection Act,*[6] and the *Canadian Environmental Assessment Act.* "These statutes deal with, among other things, the conservation and protection of fish and fish habitat, the protection of navigable waters to ensure safe navigation, and the assessment of potential adverse environmental effects associated with projects" (DFO 2006a). According to Howlett and Rayner (2004, 175), these are "classic command and control pieces of legislation" that permit DFO to play a strong role in regulating the aquaculture industry. At the same time, these statutes also restrict what DFO can do in terms of aquaculture policy. For instance, the *Canadian Environmental Assessment Act* requires that DFO lead a federal environmental assessment (EA) for aquaculture projects that involve "the construction, operation, modification, decommissioning, abandonment or other undertaking in relation to [an anchored marine structure]" (DFO 2006b). EAs are laborious and expensive projects that usually involve reviews by multiple government agencies and that can take several years to complete. Similarly, the *Fisheries Act* requires DFO to assess the potential damage of most aquaculture sites to fish habitat (DFO 2006c). This process, which is termed a Harmful Alteration, Disruption or Destruction (HADD) review, also involves a direct and extensive state evaluation of aquaculture sites and operations.

In recent years, however, DFO has been distancing itself from the command-and-control approach and, when legally permissible, minimizing its role in aquaculture regulation. Since 2000, it has released a series of policy positions and documents that downplay its powers to make demands of the industry, and make liberal use of new terms such as "creative governance approaches" and "adopting a solutions-oriented approach" to aquaculture (DFO 2002). For example, in 2000 it released its Aquaculture Action Plan, which was explicitly intended to create "an enabling regulatory environment" for the aquaculture industry (DFO 2000):

> The aquaculture industry needs regulations that will enable the industry to develop in a sustainable manner. DFO is taking action to ensure that its significant responsibilities relating to environmental protection and fish health ... are effectively administered in a transparent manner, and where appropriate [are] harmonized with provincial requirements. In this manner, DFO will work toward developing an enabling regulatory framework for the aquaculture sector – one that clearly conveys the rules for the industry and allows the sector to position itself accordingly.

The action plan identifies DFO's priorities in future aquaculture policy as "increasing industry competiveness in global markets" and "increasing public confidence in the sustainability of aquaculture development" (DFO 2000). These priorities (which address two dimensions of the triple pressure discussed earlier) have little to do with the official departmental mandate coming from the *Fisheries Act,* the *Navigable Waters Protection Act,* and the *Canadian Environmental Assessment Act.*

The Aquaculture Action Plan was soon followed by an Aquaculture Policy Framework that is intended "to inform the daily actions and decisions at DFO" with respect to the industry (DFO 2002). This document further establishes the department as an explicit enabler of industry:

DFO will make every effort to understand the needs of the aquaculture industry and to respond in a manner that is solutions oriented and supportive of aquaculture development. To help position the Canadian aquaculture sector to achieve its full developmental potential, DFO ... must be willing and able to assume an enabling role. This role demands that DFO be creative, solutions-oriented and focused on understanding the needs of, and building stronger relations with, representatives of the aquaculture sector.

This policy framework explicitly affirms DFO's recognition of the aquaculture industry as a valued DFO client group. When developing and applying operational policies, laws, programs or any other governance measure affecting aquaculture development, DFO will consider the interests of aquaculturalists and will adopt a solutions-oriented approach to dealing with specific challenges. (DFO 2002, 30)

This language is very ambitious, and some critics have argued that DFO has proven much more competent at adopting the jargon of smart regulation than at actually implementing it (Howlett and Rayner 2004). Nevertheless, there is evidence that the department is moving in two significant directions to soften its heavy regulatory hand.

First, it is moving to work more closely with the aquaculture industry by applying a strategy called "adaptive management" (DFO 2006c, 2006d). Generally speaking, adaptive management refers to "attempts to base management decisions on site-specific information" rather than adhering to firm rules or regulations that apply across the board (Blumenthal and Jannik 2000). In other words, adaptive management is about flexibility. As part of the Aquaculture Policy Framework, DFO released "interim guides" to its staff that direct them to apply adaptive management principles when dealing with aquaculture firms whose sites are likely to have an environmental impact (DFO 2006c). For instance, if a HADD review suggests that an aquaculture site is likely to damage fish habitat, DFO assessors are instructed to approve the project and negotiate a "HADD Avoidance, Mitigation, and Monitoring Agreement" with the proponent. The DFO describes these agreements as "responsive instruments ... [that] will be used as an adaptive management approach to deal with the uncertainty [surrounding environmental impacts]" (DFO 2006c). These agreements allow for development to proceed provided that the firm engages in monitoring activities and report data to DFO. (Self-monitoring is also an important provincial strategy, as we shall see.) Firms are also required to outline mitigation activities in the agreement. According to DFO, "the use of these agreements is a formal, systematic, and rigorous approach to learning from the outcomes of [aquaculture site] management actions, accommodating change, and improving [aquaculture site] management" (DFO 2006c).

A second way in which DFO is softening its regulatory hand is by deferring to provincial regulations and standards (again, when legally possible). For example, while DFO continues to lead EA and HADD reviews, it will defer to provincial pollution thresholds and standards for data collection in making its assessments, "provided the data obtained meets DFO's requirements and DFO has unrestricted access to, and use of, all of the data" (DFO 2006c). Provincial standards for key pollutants (sulfides) are now in place in New Brunswick and British Columbia.

In other cases, DFO is "reinterpreting" current laws and regulations in order to lessen its involvement in aquaculture regulation. The most significant example is found in the "Interim Guide to Fisheries Resource Use Considerations in the Evaluation of Aquaculture Site Applications," where the department states that it is deferring the licensing of aquaculture operations to provincial authorities without requiring federal approval:[7]

> In most provinces, the Provincial Government acts as the leasing authority for aquaculture operations ... Whether the [federal] Minister has the legal tools to prevent or require that his approval be obtained prior to the issuance of the lease depends on the wording in the applicable federal regulation ... Considering the wording in the relevant regulations, *DFO's approval of the provincial lease ... is not required.* (DFO 2006a; emphasis in original)

While the Minister of Fisheries and Oceans reserves the right "to be consulted on and to provide recommendations regarding the issuance or expansion of leases issued by provinces, these recommendations and/or advice will [only] be taken into account by the provincial leasing authority" (DFO 2006a).

Many provinces are also involved in the devolution of responsibilities for aquaculture policy and regulation to the industry. This is happening in two ways. First, several provinces are encouraging the aquaculture industry to take an active role in *creating aquaculture policy*. Specifically, select industry associations have been asked to develop formal codes of practice, which are then incorporated into formal government regulations and policy. Second, some provinces are implementing mechanisms for aquaculture companies to self-monitor – that is, to perform tests and collect data on behalf of regulators with respect to environmental compliance.

Since the late 1990s, aquaculture industry associations have made the development of codes of practice a major priority across the country. The British Columbia Shellfish Growers Association (BCSGA) released a Code of Practice in 2001 that includes guidelines on waste management, site density, navigation, vehicle operation, and seed collection (BCSGA 2001). Similarly, the BCSFA adopted a Code of Practice in 2005 covering issues such as feed management, net maintenance, predator control, and site appearance (BCSFA

2005e). On the Atlantic coast, the NBSGA developed an Environmental Code of Practice in 2002. The Prince Edward Island Aquaculture Alliance (PEIAA) has drafted an Environmental Code of Practice covering site installation, vehicle maintenance, and plans for chemical spills. The Newfoundland Salmonid Growers Association (NSGA) has also developed a "Code of Containment" to prevent farmed salmon escapes (OCAD 2003).

There are several motivations for development of these codes. On the one hand, the industry knows that there is a public relations advantage to the creation and publication of industry-wide standards and standard operating procedures. According to Ruth Salmon (executive director of CAIA, and formerly of BCSGA), "a Code of Practice gives an ease of mind to us dealing with retail shops that are going to want more information. Here's the product, here's where we get it from, and here's how they raise it" (Salmon 2004).

Industry-driven codes of practice are also part of the movement away from traditional command-and-control governance and towards devolution and self-regulation (Gunningham and Rees 1997). To be clear, they are not equivalent to "official" regulations. In many cases, they are wholly voluntary. In other instances, they are mandatory as a condition of membership in the industry association (this is the case with the BCSFA), although monitoring and sanctions are handled internally. What is significant about the case of aquaculture, however, is that some industry-drafted codes of practice *are being incorporated into official government regulations and policy.* For example, in New Brunswick, "compliance by growers with the NBSGA Code of Practice is tied to the licensing of salmon farming sites by the province" (OCAD 2003, 47). In other words, the government of New Brunswick requires that firms adhere to the industry code. In British Columbia, the province explicitly adopted the BCSGA's Code of Practice "as a baseline" for its own code that "is enforceable as a condition of the shellfish aquaculture license" (British Columbia 2002b, 2). Industry-driven codes are also being recognized at the federal level. For instance, a promise to adhere to Prince Edward Island's Environmental Code of Practice can facilitate environmental assessment approvals from DFO (OCAD 2003, 48).

The use of industry-produced codes of practices to formally regulate the aquaculture industry raises several questions. First, some opponents of aquaculture development view this trend as nothing less than letting aquaculturists write their own rules (e.g., Carsten 2002; Deal 2005). Second, it is important to note that codes of practice are not exhaustive. A typical code of practice will address "best practices" and set very general goals about proper conduct, but will not address details about how to achieve them. For example, the British Columbia Shellfish Code of Practice, which is now official provincial policy, offers the following guidelines for "waste management" on a shellfish farm (British Columbia 2002b, 5):

[Shellfish] farmers shall:

- Keep farms clean, safe and sanitary at all times;
- Remove garbage and debris from the site and dispose of, or recycle at appropriate facilities;
- Remove damaged or insecure equipment;
- Ensure synthetic material waste (e.g. two-stranded polypropylene rope used for string culture) generated in harvesting operations does not re-enter the marine environment at any point after harvest;
- Develop and follow a Best Management Plan for sewage and liquid waste disposal that addresses the following: provision of toilet facilities, handling and discharge of greywater, transportation and disposal of sewage and liquid waste.

These guidelines say nothing about *how* to achieve the stated directives. State adoption of codes of practice should therefore be understood as part of the movement towards "results-based" or "performance-based" regulation, where regulatory emphasis is placed on outcomes rather than processes, thereby giving firms greater discretion over day-to-day practices (Vold 2003).

The second major trend in provincial aquaculture policy is to permit firms to self-monitor in key areas. Governments justify self-monitoring by arguing that companies are able to collect much more information on their sites than state regulators could, and that this greatly enhances government's capacities to oversee industry activities (cf. Bomsel et al. 1996). The aquaculture industry strongly supports this approach, despite the fact that it often means increased costs for firms. For example, in 2000 the NBSGA and the government of New Brunswick jointly drafted and implemented a Fish Health Surveillance Program, which "moved the responsibility for regular testing and reporting of fish health data from government to the individual sea cage operator" (OCAD 2003, 47). The NBSGA claims that this program has helped aquaculturists reduce the severity of disease outbreaks (NBSGA n.d.b).

British Columbia has also moved to implement self-monitoring. In 2002, the province legislated new Finfish Aquaculture Waste Control Regulations, which require firms to monitor and report the physical conditions of their sites. This includes reporting on ocean current strength and direction, as well as sediment sampling from the ocean bottom below the site for analysis of pollutants (British Columbia 2002c). Firms are responsible for performing the baseline tests, as well as subsequent sampling and analysis. They are also responsible for drafting "best management practices" plans to "reduce waste discharge, handle spills and [fish] mortalities in a timely and appropriate manner, and [use] husbandry techniques that ... minimize impacts on wildlife" (British Columbia 2002c). On the other side of the ledger, the provincial government committed to "undertake audits and inspections to ensure that such reporting is accurate." Since 2002, the province has expanded self-monitoring to include reporting on fish health as well as monthly sampling for sea lice.

Conclusion: Canada's Eclectic Aquaculture Policy Mix

Governments in Canada are deeply committed to aquaculture development, investing huge sums of money in subsidies and in developing new policy, regulatory, and monitoring regimes for this latecoming industry. The problem is that governments are backed into several corners when it comes to aquaculture. The first corner is a legal one. Unlike land-based activities such as forestry and mining (which are provincially regulated), and unlike commercial maritime fisheries (which are federally regulated), aquaculture falls in the middle. The Canadian Constitution states that oceans are a federal responsibility while the provinces have jurisdiction over environment and commercial enterprises. As discussed earlier, DFO is taking administrative steps to soften its regulatory hand but it remains bound by constitutional law to play a strong role in aquaculture governance. This is interminably frustrating to aquaculture proponents, who see little benefit to this overlap.

The second corner involves competing complaints. The aquaculture industry regularly protests that it is over-regulated. Canadian governments clearly sympathize with this position, as both federal and provincial policy documents are filled with statements about "creating an enabling regulatory environment" for industry (DFO 2000), and "reducing the regulatory burden and enabling the growth of this important industry" (British Columbia 2002d). On the other hand, environmentalists and other opponents complain that governments are too permissive of the industry – too willing to tolerate environmental damage and unwilling to enforce existing environmental laws (particularly the federal *Fisheries Act*) (Langer 2004). Although governments have largely ignored the latter claims, they have taken a toll on public acceptance of the industry and of government attempts to regulate it.

The third corner in which governments are stuck is the most serious. In order for aquaculture to work in Canada, governments must address three problems: the global competitiveness of the industry, its long-term environmental integrity, and the need to legitimize both the industry and government policies towards it. In this chapter, we saw that federal and provincial governments are using both traditional and innovative policies to address this triple pressure. On the one hand, Canadian governments have gone down the well-travelled road of extensively subsidizing aquaculture development. These subsidies are primarily intended to address the competitiveness problem. Many subsidy programs are aimed at encouraging investment in Canadian aquaculture by sponsoring industry expansion, while others are intended to increase productivity by supporting research and development and technological uptake. We have argued, however, that subsidies do little to address the other dimensions of the triple pressure (except the fact that governments are also subsidizing research on environmental issues), and may in fact harm efforts to legitimize the industry.

On the other hand, some governments in Canada are responding to the triple pressure in non-traditional or innovative ways that involve "going beyond government" in two respects. First, governments have initiated legitimacy-building activities that engage with stakeholders and outside experts, although such efforts have a mixed record. Some Atlantic governments have been able to establish permanent (but limited) means for community and stakeholder engagement (such as Nova Scotia's Regional Aquaculture Development Advisory Committees). On the Pacific coast, however, attempts to establish permanent legitimacy-building activities have failed, either because the process did not work very well (as with Community Shellfish Steering Committees) or because government itself chose to make an end-run around stakeholder groups in arriving at key decisions (as with the Salmon Aquaculture Implementation Advisory Committee). The second non-traditional policy response involves the devolution of authority and key responsibilities to industry. According to supporters of this strategy, devolution is a way of both increasing competitiveness and improving environmental compliance. At the federal level, DFO has committed to working with aquaculture firms using a strategy called adaptive management in order to come up with site-specific plans to mitigate environmental impacts. At the provincial level, industry associations have been encouraged to develop codes of practice, which are then (in some provinces) incorporated into official state policy. Several provinces have also actively promoted industry self-monitoring on key environmental criteria. Advocates of this approach argue that self-monitoring will improve environmental performance over the long term by increasing volumes of site-specific data. Critics charge, however, that devolution and self-monitoring give the industry too much influence over the regulatory process and weaken state authority over its primary mandate: the protection of the environment and public goods.

It is still too early to tell whether the Canadian approach to governing aquaculture will enable the industry to survive and expand. What is clear is that aquaculture poses a serious challenge to government policies and practices. In the past, governments would have made and implemented their decisions on aquaculture according to their own priorities for the industry. According to Howlett (2001, 303), governments are much more sophisticated now than in the command-and-control past but, paradoxically, "their autonomy or ability to effect change independently have been eroded." The result, for the moment, is a patchwork of policies, each of which is controversial in its own right. Moreover, it is difficult to get a sense of where aquaculture policy is going in Canada. On the one hand, there are clear trends. For instance, the Department of Fisheries and Oceans appears committed to softening its regulatory hand in aquaculture. At the provincial level, the trend towards devolution appears equally strong, as New Brunswick and British Columbia in particular are implementing more and more mechanisms for industry self-regulation and self-monitoring. On the other hand, governments continue to roll out old-fashioned subsidy programs for aquaculture development, despite the fact that these delegitimize the industry in the eyes of many stakeholders. The challenges facing Canadian aquaculture are significant, and governments are emptying the policy toolbox in their attempts to secure and stabilize the industry's future.

Conclusion

As stated in the Introduction, this book is about the *aquaculture controversy*, not about the pros and cons of the industry or the desirability of aquaculture development. There are many voices arguing for and against the expansion, reform, and regulation of salmon, freshwater, and shellfish aquaculture in Canada. Our task has been to analyze these disagreements (rather than join them) and examine *why* the controversy over aquaculture exists and *how* it has been perpetuated. As we have seen, ongoing conflicts over aquaculture represent one of the most divisive and intense struggles over industrial development ever to have taken place in Canada. Although much of this debate revolves around specific disagreements over the immediate consequences and benefits of aquaculture, we have argued that the conflict is rooted in deeper forces and issues that give it its particular acrimony and longevity.

We have argued that at its most basic, the aquaculture controversy is about more than it seems. Our thesis has been that the conflict is both unique (reflecting the specific history of coastal and resource development in Canada) and rooted in some of the major unresolved questions about environment, knowledge, governance, rights, and development that are confronting democratic societies around the world. As such, a full explanation of why this controversy has arisen and endured requires that we look at the broader context, both of aquaculture's evolution as a major coastal industry and of the different scientific and moral positions that anchor claims making on all sides of the debate.

The Late Emergence of Aquaculture in Canada
As discussed in Chapter 1, aquaculture is one of the few genuinely new resource industries to emerge in Canada during the past half-century. We argued that this late arrival has profound implications for the industry and the controversy in three main ways.

First, aquaculture's late emergence in Canada means that it is entering the global marketplace from a position of considerable weakness. Canadian resource industries have enjoyed a historical advantage in the international market due in large part to the vast spaces and high volume of production in this country. Until very recently, Canadian forestry, mining, and fisheries had a strong influence on prices for key commodities because of their dominant market position, particularly in exports to the United States. We saw in Chapter 1, however, that this dominance began to recede with the acceleration of economic globalization and the entry of new producing regions possessing substantial cost advantages. While traditional industries (particularly forestry) are struggling to adapt to these changes, the late emergence of aquaculture means that it is dealing with globalization as a small producer and minor international player. With respect to salmon aquaculture in particular (the most lucrative sector in Canadian and world aquaculture), other nations are dominant on both the high road (Norway) and the low road (Chile) of aquaculture production. As a result, Canada's aquaculture industry has followed the well-travelled route of taking advantage of proximity to the United States. This reliance on lower transportation costs for competitiveness means that Canadian aquaculture has become yet another industry in this country whose profitability is based on moving high volumes of minimally processed products to a single export market. This is a fragile position in the highly competitive North American marketplace, and it means that additional costs (from labour, from regulatory compliance, or incurred through protest and opposition) directly threaten the industry's viability.

The second consequence of aquaculture's late emergence is that it is forced to deal with the legacies of other industries. Proponents of aquaculture see the industry as a legitimate user of ocean spaces on par with existing activities such as fisheries, transportation, tourism, and energy (on the Atlantic coast). Aquaculture, however, is entering spaces that are already severely degraded because of prior and existing use. Canada's nearshore spaces have been abused not only by ocean industries but also by forestry (stream degradation and mill effluent), mining (tailings and run-off), construction, property development, utilities, and waste management industries. As we saw in Chapter 2, aquaculture's late claims to ocean spaces means that many opponents see this as a make-or-break issue. Aquaculture is not responsible for pushing ocean ecosystems to the brink, but many opposition groups and local stakeholders clearly fear that it will push them over. Importantly, aquaculture's late emergence means that it confronts the *environmental legacy* of other industries without a *social or cultural legacy* of its own. The data in Chapter 2 showed that many local people are still committed to traditional industries such as fisheries and forestry because of their link to coastal ways of life or livelihood. As a new industry, aquaculture's environmental impacts are not culturally justifiable in this way.

The third consequence of aquaculture's late emergence is that governments are still figuring out how to regulate the industry. As discussed in Chapter 7, both federal and provincial governments are still experimenting with aquaculture policy and regulations. Aquaculture challenges governments because of its economic fragility, presence in environmentally sensitive spaces, and problems of public legitimacy (what we termed the "triple pressure"). Existing fisheries and agricultural legislation have proven to be a poor fit, and the industry regularly complains that jurisdictional overlap places an unnecessary regulatory burden on aquaculture firms. On the other hand, critics strongly object to the aggressive subsidizing of the industry (necessary for its economic survival) and argue that governments are willfully sacrificing their environmental protection mandates in order to promote aquaculture development. Although new policy tools are emerging with the goal of building legitimacy and enhancing industry flexibility, these are untested and uneven, and threaten to deepen the controversy by giving significant leverage to industry to both self-monitor and play a key role in developing aquaculture policy on sensitive issues such as pollution (see Chapter 7).

Local/Global Tensions
The aquaculture controversy in Canada is also being shaped by significant tensions between local and global issues and actors.

We argued in Parts 1 and 2 that anti-aquaculture groups have had significant success in linking local or regional issues with national- and global-scale activism. As seen in Chapter 1, the global environmental movement now plays a significant role in debates over industrial development across the democratic world. Much of this movement's power comes from the successful assertion that environmental problems, even when they occur in specific places, are a legitimate concern for everyone. The logic of global environmentalism (which differs from the preservationist and conservationist movements of the early twentieth century) opens the door for stakeholders across Canada and the world to legitimately advance concerns and claims about a regional industry. This means that many of the battles over the future of the industry take place far away from the rural coast, in activist networks, expert publications, media rooms, laboratories, and websites. It also means that activism comes from multiple sources and scales at once, which poses a significant strategic and communications challenge to the industry and its supporters. Returning to the metaphor introduced in the Introduction, this complexity means that the aquaculture industry is caught in a labyrinth (Leiss 2001), forced into short-term reactions to unforeseen twists and turns. (We also argued in Chapter 3, however, that the industry has failed to anticipate and respond to some aspects of the controversy despite ample warning.)

Local/global political and economic tensions are also shaping the aquaculture controversy. As resource-producing regions, both Atlantic and Pacific coastal communities are accustomed to seeing the extraction and export of local natural capital. At times, these arrangements have brought prosperity and stability to coastal regions, whereas at other times they have done the opposite. As we saw in Chapter 2, exploitation and dependency are sensitive local issues, particularly in the current extended period of economic hardship on Canada's coasts. Local opinion differs on how aquaculture fits in with these experiences, however. Some coastal stakeholders see aquaculture within what we termed a "continuity narrative," where industry investment and employment can help perpetuate coastal lifestyles and livelihoods. Others use a "disruption narrative," seeing aquaculture as an extension of corporate control over local resources at the expense of locally or at least regionally based commercial fisheries. These divisions are even more profound in and among Aboriginal communities, where some see involvement in aquaculture as a means of asserting rights claims, while others view it as a neocolonial imposition by governments and corporations on traditional rights and territories.

The passion with which these different views are articulated (see Chapter 2) leads us to argue that aquaculture has become a powerful but fluid metaphor in the local context, representing a myriad of hopes and fears under very stressful economic and environmental circumstances. This metaphorical status has been encouraged by some industry supporters who argue that aquaculture is a kind of saviour for coastal communities struggling under economic hardship (a position that may win the industry support in some quarters but probably also galvanizes opponents who are not willing to concede that traditional activities are lost). In our view, the status of aquaculture-as-metaphor is a significant factor in the perpetuation of the controversy. Despite their opposing stances, both supporters and opponents of aquaculture at the local level have associated the industry with the preservation of livelihood and a way of life. It is difficult to find a middle ground when the industry has come to represent similar but opposing hopes and fears.

The Changing Politics of Knowledge

We have also argued that changes in the politics of knowledge have permitted the aquaculture controversy to emerge, go public, and endure over time. As we saw in Part 2, the public struggle between pro- and anti-aquaculture groups is predominantly a knowledge conflict, as arguments about the effects and consequences of aquaculture are based primarily on scientific and expert claims. Scientific knowledge is accorded significant authority in Western cultures, but we argued in Chapter 3 that scientific and expert voices have become more plural in recent years, in a manner that has opened up room for "alternative sciences" that challenge the claims of traditional scientific authorities, particularly in industry and government.

These changes, which include the rise to prominence of ecology as a "subversive" science, the institutional diffusion of scientific skills, and pressures to engage in a more "inclusive" science that is transparent and open to public inputs, have enabled *multiple science-based narratives about aquaculture to coexist at the same time over the long term.* On the one hand, the diffusion of scientific capacity to non-traditional institutions (such as environmentalist groups), along with the expansion of environmental science (which is based on the collection and modelling of large data sets, both of which are becoming easier with technological advances), means that the production of alternative scientific claims is more possible today than twenty years ago. On the other hand, the push for a more inclusive science, along with new capacities for stakeholders to access information and become "experts" themselves on complicated issues, means that concerned members of the public are more willing to challenge official science and consider alternative claims. Thus, more scientific investigation, which we would normally expect to lead to the clarification of aquaculture issues, can in fact contribute to less clarity and consensus (see Chapter 3).

Opponents have taken advantage of these openings to mount forceful challenges to official government and industry claims about aquaculture. We saw in Chapters 3 and 5 that environmental nongovernmental organizations, in particular, have demonstrated significant skill in blending scientific claims with cultural narratives. Our overview of pro- and anti-aquaculture attempts at strategic framing showed that oppositional narratives of "taint" and "risk" have had strong traction in media coverage of the industry, particularly in the national and Pacific region. This dominance is such that when industry supporters are quoted in the media, they are disproportionately likely to be defending the industry against claims of risk and harm rather than advancing their own narratives about the benefits of aquaculture (see Chapter 5). Moreover, we found that opposition groups have been successful in establishing and maintaining an offensive stance on the "knowledge battlefield" through strategic publications and press releases. On pivotal issues such as pollution, escapes, human health, and sea lice, opponents have seized the communications initiative, enabling them to choose points of attack, selectively engage with pro-aquaculture claims, and establish a powerful weight-of-evidence narrative and broad moral stance (see Chapter 3). By ceding the communications initiative, industry supporters almost always find themselves in a reactive or defensive stance when these issues (repeatedly) grab the public's attention. Although the industry mobilizes significant scientific resources to defend itself, we argued that its point-by-point rebuttals (which tend to contain a much narrower moral narrative focusing on bias or wrongs to the industry) are not good vehicles for building credibility in the broader public sphere.

Actions on the knowledge battlefield are also having an impact on scientists and experts themselves, which in turn pushes the debate further away from resolution or consensus. We saw in Chapter 4 that there is plenty of room for expert disagreement on aquaculture issues, particularly on questions of environmental impacts. While disagreement is normal in the natural sciences, our survey of aquaculture experts in Canada found markedly high levels of distrust among experts with different institutional affiliations and stances on aquaculture. Without trust, scientists have difficulty building on one another's findings and addressing one another's critiques and concerns. In the case of aquaculture, we argued that distrust makes "knowledge control" a key issue: despite the good intentions of many aquaculture experts, most are resistant to fully engaging with adversaries, stakeholders, and "the public" (i.e., the general public and institutions such as the media) in producing and managing knowledge.

Conclusion: The Labyrinth

The aquaculture controversy in Canada has changed over the years since its eruption in the 1980s, but it has also remained markedly consistent. While new claims and issues are constantly emerging, the main axes of debate – environment, human health, rights, and development – have endured. In our view, this persistence is proof that the aquaculture debate is about more than just the rise and expansion of a new industry. Rather, the controversy is rooted in the specific challenges facing coastal Canada (economic and environmental), and in broader unresolved questions about the limits of industrial development; the validity of competing knowledge claims; the rights of nature, stakeholders, and indigenous peoples; the capacity of democratic governments to balance interests and responsibly protect the environment; and the future of peripheral places in an increasingly globalized world.

This complexity means that aquaculture proponents need to be farsighted. Economic, environmental, and political pressures are pushing governments to experiment with different policy mixes, and are pushing the industry into a reactive and defensive stance that more often than not involves the assertion that worries and criticisms are unfounded. The complexity of the debate means that the long-term implications of these actions are not always clear. This is the essence of Leiss' metaphor (2001) of the labyrinth: dealing with problems as they arise may seem like a pragmatic approach to problem solving in the face of controversy, but this kind of twisting and turning will not address the complex roots of the conflict. Taking the metaphor further, we suggest that the attempts of government and industry to escape the labyrinth as quickly as possible – to defuse the controversy and get on with business – may in fact be leading them deeper into the maze. Throughout this book, we have argued that aquaculture *itself* has become a powerful but very fluid metaphor. Because this industry has emerged at a time of profound change – both global and local, economic and environmental, political and cultural – aquaculture has come to represent complicated *hopes and fears* about the future. Very few industries find themselves at the intersection of so many larger issues. For Canada's aquaculture industry to survive, it must recognize the validity of these concerns even if it disagrees with them, and find ways to address criticisms over the long term. As we saw in Chapter 3, lessons from earlier industrial controversies show that transparency and a willingness to engage with stakeholders are crucial in establishing and maintaining public trust in times of conflict. While it remains to be seen, this may be the only path through the labyrinth.

Notes

Introduction

1 Quotes are from Edgar 2002 and Meissner 2002.
2 Hicks 2004; Mittelstaedt 2004; Hites et al. 2004.
3 Quotes are from Zacharias 2005.
4 Salmon Farm Monitor 2005.
5 Simpson 2005.

Chapter 1: Aquaculture in a Global Context

1 *Delgamuukw v. British Columbia*, [1997] 3 S.C.R. 1010.
2 At the time of writing, the average price remains low at US$4.16/kg (February 2009).
3 Jessop (2002a, 55-59) uses the label "Atlantic Fordism" to describe the dominant postwar economic system, and the term "Keynesian Welfare National State" to describe the state form that accompanied this system. In the interests of clarity, and given that our aim here is less theoretical and more analytical, we have chosen to simplify these labels and employ the more commonly used term Fordist-Keynesian economy and state.
4 This, of course, assumes that local actors have the capacity to pursue these goals. Recent research suggests that empowerment without capacity can create rather than assuage regional inequalities (Brenner 2003; McCarthy 2006; Young 2008).
5 We are not alone in making this suggestion. For other perspectives on the neoliberal "compromise" in resource regions, see Herbert-Cheshire 2003; Herbert-Cheshire and Higgins 2004; McCarthy 2005, 2006; and Beer et al. 2005.

Chapter 2: Aquaculture in a Local Context

1 The Special Committee on Sustainable Aquaculture was composed of ten Members of the Legislative Assembly (MLAs) of British Columbia. It was a unique committee in that a member of the opposition New Democratic Party (NDP) was appointed chair and (centre-left) NDP votes outnumbered those of the ruling (centre-right) BC Liberal Party.
2 Because these proceedings are part of the permanent public record of the government of British Columbia, we identify the speakers of the quotations given in this chapter.
3 These are not the only bases of local opinion. Local understandings of aquaculture draw on both scientific and local knowledge and experiences. Many coastal residents are highly aware of ongoing expert and political debates regarding the environmental and health effects of aquaculture production (see Chapter 3).
4 *Calder v. British Columbia (Attorney General)*, [1973] S.C.R. 313, [1973] 4 W.W.R. 1; *R. v. Sparrow*, [1990] 1 S.C.R. 1075; *Delgamuukw v. British Columbia*, [1997] 3 S.C.R. 1010.
5 *R. v. Marshall*, [1999] 3 S.C.R. 456.

Chapter 3: Knowledge Battlefield

1 While this ideology is predominant, it is not uncontested. The dark side of science and technology has always had a place in modern culture (and is most saliently expressed in works of dystopic science fiction, from H.G. Wells to Michael Crichton; cf. Moylan 2000).

2 Strictly speaking, ecology and ecosystem science do engage in reductionist analysis (McIntosh 1987). That is, ecology often "seeks explanation for phenomena by looking to smaller scales than those at which the observations were made" (Weigert 1988, 269). This is a key source of knowledge and theorization in ecology, but even the suggestion of reductionism grates with some ecologists, and debates continue within ecological sciences about the place of "reduction versus holism" (see Levins and Lewontin 1980; Wilson 1988).

3 This conceptualization of master frames differs slightly from that advanced by Benford and Snow (2000) and Carroll and Ratner (1996), who understand master frames as the overarching ideological frame through which a social movement understands its actions (such as "rights frames," "injustice frames," "environmental justice frames," etc.). Our conceptualization of master frames draws primarily on the literature concerned with framing as a strategy for communication rather than as a fundamental motivation for action (cf. Gamson and Modigliani 1989; Dryzek 2005; Hannigan 2006), although we note that these differing conceptualizations are not incommensurable (see Stefanik 2001).

4 http://www.farmedanddangerous.org. The image was accessed in June 2007.

5 POSA members in 2005 were: Creative Salmon Co. Ltd., Ewos, Nelson Island Oysters, Omega Pacific, Pacific Seafoods Ltd., Seaspring Hatchery, Taplow Feeds, Totem Sea Products Ltd., West Coast Fish Culture, and Yellow Island Aquaculture.

6 The dailies are the *Globe and Mail, National Post, Vancouver Sun, Vancouver Province, Victoria Times Colonist, Halifax Daily News,* and *St. John's Telegram.* The date range of articles examined from individual dailies varies. Please see Chapter 5 for a detailed discussion of the methodology and limitations of the study.

7 *Framed Salmon* was originally written as a rebuttal of the earlier PCB studies (Easton and Luszniak 2002; Jacobs et al. 2002a, 2002b) but was updated following the publication of the Hites article. The new version was widely distributed as a quick response to the 2004 crisis.

Chapter 4: Knowledge Warriors?

1 We sincerely thank all who participated in the survey. This research would not have been possible without their time and candour.

2 Students were excluded from this sample, mostly because we were often unable to assess their area of study or progress in their program (e.g., beginning or near completion of a graduate degree) from the information available.

3 The survey has a unique design. In general, the strength of surveys lies in their ability to generate large amounts of comparable data. Among their weaknesses, however, is the fact that they are relatively inflexible and incapable of capturing nuances and deeper meanings (Gray and Guppy 1999). To minimize this weakness, we conducted the survey through the Internet (participants were invited by letter to visit a secure website and enter a personal identification number). Given the high technical literacy of our participants, we did not expect Internet access to be a major barrier to participation. The electronic format of the survey had the advantage of allowing for multiple open-ended questions and opportunities for respondents to comment directly on the issues raised. In a traditional paper format, this would have made the questionnaire forbiddingly long. The electronic version allowed participants to comment when desired and at length. Thus, we are able to combine findings from both closed and open-ended data in our analysis.

4 The totals exceed 100 percent due to rounding. Of the 287 valid responses to our educational achievement question, 171 respondents indicated PhD, MD, DVM, or equivalents (several respondents held advanced law degrees) as the highest credential attained; 57 indicated MSc, 48 BSc, and 11 Other.

5 Responses of "don't know" were excluded.

6 A score of 0 could also indicate neutrality or uncertainty (if many respondents chose the option "neutral or undecided"). The standard deviation of all items near 0 (between –0.5

and +0.5) is high (exceeding 0.75), indicating that averages near 0 are being driven by disagreement rather than neutrality or uncertainty.

7   In each case, the standard deviation exceeded 0.5, indicating significant variance in responses despite the relatively high means.

8   All survey research is vulnerable to deliberate attempts to provide misinformation. Given that we did not expect many positive evaluations of aquaculture from experts affiliated with environmentalist groups, we looked very carefully at the outlier surveys. In our judgment, they are valid and are representative of the diversity of opinions among groups that self-identify as environmentalist organizations. The outliers are included in all of our subsequent statistical analyses.

9   The *eta*-squared value is significant at < .01.

10 Eleven of fifteen ENGO-affiliated experts responded to all questions used to measure individual situation. Given all possible permutations of previous employment and of scores on the collaboration index, we deem this too small a population for this analysis. Several other factors discourage us from pursuing this line of inquiry with respect to environmentalist groups. As shown in Table 4.2, the ENGO category has a very high standard deviation (9.6). This is due to several outliers on the upper range of the Stance Scale. Given the small sample, the individual situations of the experts giving these outlying responses will show up as statistically significant only because of their marked difference from the small population of other respondents.

11 For the regression analyses, we convert the −14 to +14 scale to a 0 to 28 scale. The multivariate regression model equation for the prediction of an industry-affiliated respondent's stance on aquaculture is $y = -5.64$ (female) + 4.81 (previous employment with government) + 22.75 (constant). Thus, female gender is associated with a reduction in score on the Stance Scale of 5.6 (from a constant of 22.75). This is a large effect, given that the standard deviation for industry-affiliated experts is only 4.4 (see Table 4.2).

12 The multivariate regression model equation for the prediction of a university- or college-affiliated respondent's stance on aquaculture (on the 0 to 28 scale) is $y = 7.07$ (previous employment with industry) + 1.48 (collaboration activity with industry colleagues) − 1.64 (collaboration activity with ENGO colleagues) + 13.59 (constant).

13 Step 2 of the regression model considers the institutional and individual variables together. This means that the explanatory power of the institutional variables is recalculated in light of the individual variables, rather than their being simply summed. This is a strength of regression analysis, and leads to a very accurate accounting of which variables are explaining observed variance. Thus, the 28 percent explained variance in Step 2 comes from the adjusted R^2 value from the multivariate regression model: $y = -5.82$ (university or college affiliation) − 12.6 (ENGO affiliation) + 6.94 (university or college affiliation, previous employment in industry) + 11.33 (ENGO affiliation, previous employment in university or college) + 1.91 (university affiliation, collaboration activity with industry colleagues) + 1.71 (industry affiliation, collaboration activity with industry colleagues) + 18.99 (constant).

14 Based on *F*-test of analysis of variance. This test compares variance within groups with variance between groups. Thus, insignificance here means that differences of opinion across supporters, moderates, and opponents are not sufficiently distinct from differences in opinion within the groups.

15 The difference in means across these categories is significant at < .01 ($F = 6.6$).

16 At each step in the model, new decisions regarding the inclusion or exclusion of variables were made. This was done to protect against an overly inflated final percentage of explained variance. For instance, if one of the individual situation measures had a similar association with stance on aquaculture as one of the values measures, the *least significant* variable is dropped from the later step *despite* the fact that each variable is statistically significant when standing alone. This guards against the tendency for complex models (with many variables) to yield an artificially inflated R^2 or adjusted R^2 value.

17 The findings presented in this section are considered more thoroughly in Young and Matthews 2007a.

18 The quotations used in this section are taken from answers to the following open-ended questions and invitations to comment: "What factors do you think ought to guide

government policy on aquaculture?"; "We would like to give you the opportunity to comment further on your general views of [aquaculture] science and knowledge and/or any of the questions asked in this section?"; and "What do you think about the role of Aboriginal people in aquaculture?" The first two questions were posed *before* the closed statements regarding local knowledge and Aboriginal concerns. The last question was posed after respondents had addressed these statements.

19 The commentary for this section was given in response to two open-ended questions: "Are scientists doing an adequate job communicating the risks and benefits of aquaculture to the public? Why?" and "Do you believe that your understanding of the benefits and risks of aquaculture is different from the understanding held by the public?"

Chapter 5: Media and the Knowledge Battlefield

1 We were restricted in our choice of media outlets to those that were electronically archived and thus available for keyword searches. This leads to some imbalances in the newspapers selected. For instance, we have no newspapers from New Brunswick, which is a major home for Canadian aquaculture (at the time of research, electronic archives from the *Fredericton Daily Gleaner* were not available to us, while the *Saint John Telegraph Journal* was available only from 2006). This is an admitted limitation of the study.

2 Although the discussion here focuses on the *Vancouver Sun* (because this newspaper has the longest electronic record), the trends for 1999 and 2003 apply equally to *The Province* and the *Victoria Times Colonist.*

3 Each of the 1,558 items was coded by a single research assistant (Mary Liston, who is co-author of this chapter).

4 These figures sum to more than 1,558 because some articles explicitly mention both finfish and shellfish aquaculture.

5 This number appears low, particularly since environmentalist groups have long been championing closed-containment systems as the best way forward for the aquaculture industry. The closed-containment argument is most often made within an ecological risk or harm framework, however (i.e., as a way to reduce environmental damage). We applied the "economic growth through environmental reform" code only when the argument was made that industry could improve its bottom line and/or business practices by being more "environmentally responsible" in this or other ways.

6 Judging whether or not an organization is oppositional is not always clear. We draw the distinction on the basis of whether or not the organization has taken a public position against some form of aquaculture. For example, the United Fishermen and Allied Workers' Union in British Columbia has publicly stated its opposition to salmon aquaculture, so we have included it in this category. In contrast, at the time of writing, the Fish, Food and Allied Workers Union of Newfoundland and Labrador had not taken a public position against aquaculture (to our knowledge), meaning that it is excluded from this category.

7 As before, all ratios reported here are significant to 0.01 at the .01 level, or 99% (*phi*).

Chapter 6: Aquaculture and Community Development

1 This study would not have been possible without the work of Murray Shaw and Erika Paradis, both of whom were graduate students at the University of British Columbia at the time of research.

2 In order, the top ten are oil and gas extraction, automobile manufacturing, metal smelting and refining (excluding aluminum), petroleum refining, aerospace manufacturing, aluminum production and processing, paper milling, sawmilling, pulp milling, and synthetic rubber manufacturing.

3 Some processors are involved in both commercial fisheries and aquaculture. Processors were included only if they were exclusively engaged in processing aquaculture products.

4 The population of direct aquaculture firms was defined by compiling provincial and territorial lists of aquaculture licence holders. Although we anticipated that there would be approximately 500 licensed aquaculture firms in Canada, this method yielded a population of 1,565 separate firms. Further research revealed that many firms holding aquaculture

licences were not involved in aquaculture at all. For example, in some jurisdictions, fish importers and even restaurants that display live fish and shellfish are required to hold an aquaculture licence. Using this information, we estimated a total of 454 active aquaculture firms in Canada (see Matthews 2004 for a complete discussion of methods used in this study).

5 One respondent noted that winter production is particularly important in salmon aquaculture, as this is an off-season for commercial fisheries, meaning that aquaculture dominates the fresh salmon market at this time. Most of the firms employing 100 or more workers are involved in salmon aquaculture.

6 The estimates given in Table 6.2 were achieved using slightly different methods. The most straightforward method was used for firms with 5 to 24 employees, 25 to 49 employees, and 50 to 99 employees. In these cases, we multiplied the estimated number of firms in each category (provided by CAIA) by the mean number of employees discovered through the survey. For example, CAIA identified 127 firms in the population sample as being in the 5-to-24-employees category. We multiplied this by the mean for last week (9.65), average week (9.30), and busiest week (9.45) to arrive at the figures given in Table 6.2 (1,226, 1,181, and 1,200).

The estimates for the smallest and largest firms were done differently. The category 0 to 4 employees presents a unique problem for our estimates. Our survey measured employment by asking the question "Thinking of an average / last / busiest week, how many male and female employees were on your company's payroll in each of the following types of employment?" Many small operations are family-based firms where multiple members of the family tend the operation on a part-time or full-time basis. It would appear that a large number of the operators of such firms do not see themselves and their family members as being paid by the company, and therefore answered "zero" to this question. Clearly, at least one person is at work in these operations, and probably more than one if these are family operations. In deciding how to incorporate this into our estimates, we compared questionnaires in this category with those that indicated 1 to 4 employees. There appear to be no substantive differences between the two groups on other variables (such as production, activities, and business history). Therefore, for the purposes of estimating total employment, we combine zero-employee firms with those having 1 to 4 employees, and assume that the mean for the latter is valid for all firms in this new category (2.4 persons per firm). This has the potential for introducing error, but it is a reasonable assumption to make, given that these firms certainly have at least one person working the enterprise, and probably at least one family member assisting. Finally, CAIA was unable to estimate the number of firms with fewer than 5 employees. Note 4 described how we reached an estimate of 454 direct aquaculture firms in Canada, and CAIA's information shows that 157 of these have 5 or more employees. Hence, it follows that 297 fall into the residual category. This figure, multiplied by the average (2.4 persons), gives us the totals of 710 workers during the last week, 704 during the average week, and 725 workers during the busiest period.

The other problematic category for our estimate involves the largest firms, those with 100 or more employees. CAIA estimated that there were 6 firms in this category. Our survey, however, found 11 for the last week and average week, and 12 for the busiest week. This discrepancy can be partially explained by looking at how firms chose to submit information. For instance, one firm chose to submit separate questionnaires for Atlantic and Pacific regions. In other cases, CAIA identified an overarching or umbrella company, whereas the questionnaires were in fact completed by multiple subsidiaries. Given our general knowledge of the industry, we were able to discern that we were in fact missing data from only one major employer in Canada. We therefore estimate that firm's employment based on the average number of employees of other firms in this category. This produces a total estimated employment among the largest firms of 2,280 for last week, 2,236 for an average week, and 2,679 during the busiest week.

7 It is possible that aquaculture employment has grown since our study to reflect the more recent claims of the NBSGA and BCSFA. We note, however, that the DFO estimate, which is significantly lower than those of the NBSFA and BCSFA, was published in 2005. Moreover,

available data suggest that employment in aquaculture grows very slowly, particularly in salmon farming, as automation in feeding, site maintenance, and processing continues to advance (BC Statistics 2002, 35).

8 This notion of an owner-operated commercial fishery may itself be already outdated and overly romanticized (Ecotrust Canada 2004).

9 This is a rough but conservative estimate. It is achieved by using the midpoints of each category in Table 6.8 and substituting $12,000 for the "under $15,000" category, and $80,000 for the "over $75,000" category.

10 All figures are from Community Profiles published by Statistics Canada from the 2001 census.

11 A small number of firms indicated that they had both shellfish and finfish operations. In these cases, it was not possible to distinguish between employees engaged in shellfish production and employees engaged in finfish production. The earnings of employees of these firms were therefore incorporated into both categories.

12 Respondents were asked to indicate whether they strongly agree, agree, neither agree nor disagree, disagree, or strongly disagree with each of the statements given (the categories of strongly agree/agree and strongly disagree/disagree have been combined in this table).

13 The population was determined using a list of current business licences held in the community. On close examination, we discovered that this included some businesses that were not in fact operating in Port Hardy. We therefore adopted more specific criteria: to be included in the study, a business had to have an active presence in the town or its close surroundings. Specifically, the firm had to have a place of business (an office or fixed address of some kind) in the community. This criteria encompassed all stand-alone businesses, branches or chains, and home-based businesses. Some self-employed persons were included if they held a business licence, but many were clearly not captured by this method. Surveys were administered in-person, meaning that trained interviewers asked respondents questions directly and recorded responses. This method led to a higher response rate than would be expected with a mailout questionnaire (65 percent).

14 The study was part of a Community-University Research Alliance initiative called the Coastal Communities Project, funded by the Social Sciences and Humanities Research Council of Canada. For more information, see http://www.coastalcommunitiesproject.ca.

15 Respondents were asked to "check all that apply." This means that a business could indicate involvement in multiple sectors.

16 Identical questions were also asked regarding "changes in forestry," "changes in commercial fisheries," "changes in sport fisheries," "the closing of the Island Copper Mine," "shut-downs at the Port Alice pulpmill," "the closing of Government offices," and "the sinking of the *Queen of the North*" (a ferry, operated by BC Ferries, running seasonally from Port Hardy to Prince Rupert that sank on 22 March 2006).

Chapter 7: Governing Aquaculture

1 *The Constitution Act, 1867*(U.K.), 30 & 31 Victoria, c. 3.

2 *Canadian Environmental Assessment Act,* S.C. 1992, c. 37.

3 *Fisheries Act,* R.S.C. 1985, c. F-14.

4 It should be noted that both federal and provincial governments have been forced to develop limited procedural instruments to meet their legal obligations to consult First Nations groups concerning new aquaculture projects. As discussed in Chapter 2, a series of court decisions has entrenched the rights of Aboriginal groups to be meaningfully consulted on development projects in their traditional territories. Prior to the 2004 *Haida Nation/Taku River* decision (McNeil 2005) from the Supreme Court of Canada, however, provincial governments in particular often intentionally took a backseat in consultations with First Nations groups, encouraging the project proponents (usually private companies) to take the lead. The Supreme Court ended this practice by asserting that responsibility for consultation ultimately lies with provincial governments. To date, however, provincial consultations with First Nations groups with respect to aquaculture have proceeded on a case-by-case basis, meaning that there is very little consistency or standard procedure to this procedural instrument. Consultation at the federal level is generally more streamlined,

but does not deal with specific aquaculture sites or projects. For instance, DFO has policies in place to consult with Aboriginal groups (and other stakeholders) on major legislation and policy changes (such as proposed amendments to the *Fisheries Act* and the Atlantic Fisheries Policy Review), but consultation on specific aquaculture projects is left to the provinces.

5 Predating this was the Gillespie Review, conducted in 1986, which was limited in scope and did not involve public or expert hearings (Keller and Leslie 1996, 56).

6 *Navigable Waters Protection Act,* R.S.C. 1985, c. N-22.

7 There are exceptions, as when aquaculture operations will be located on federal property "such as Port Authorities, National Parks, and for operations located offshore," and for non–Atlantic salmon species in Newfoundland and Labrador (DFO 2006a).

References

ACOA (Atlantic Canada Opportunities Agency) 2008. Aquaculture research gets major boost. News release, 1 March.

Adam, B. 1998. *Timescapes of modernity: The environment and invisible hazards.* New York: Routledge.

Ali, H. 1997. Trust, risk and the public: The case of the Guelph landfill site. *Canadian Journal of Sociology* 22 (4): 481-504.

–. 1999. The search for a landfill site in the risk society. *Canadian Review of Sociology and Anthropology* 36 (1): 1-19.

Akyeampong, E. 2007. Canada's unemployment mosaic, 2000 to 2006. *Perspectives* (January): 5-12. Catalogue no. 75-001-XIE. Ottawa: Statistics Canada.

Allan, S., B. Adam, and C. Carter. 2001. The media politics of environmental risk. In *Environmental risks and the media,* ed. S. Allan, B. Adam, and C. Carter, 1-26. New York: Routledge.

Anand, S., R. Yusuf, A. Jacobs, Q. Davis, H. Yi, P. Gerstein, and E. Montague. 2001. Risk factors, atherosclerosis, and cardiovascular disease among Aboriginal people in Canada: The study of health assessment and risk evaluation in Aboriginal peoples. *The Lancet* 358 (9288): 1147-53.

Anderson, A. 1993. Source-media relations: The production of the environmental agenda. In *The mass media and environmental issues,* ed. A. Hansen, 134-49. New York: Leicester University Press.

Apedaile, P. 2004. The new rural economy. In *Building for success: Explorations of rural community and rural development,* ed. G. Halseth and R. Halseth, 111-34. Brandon, MB: Canadian Rural Revitalization Foundation.

Aquaculture Association of Canada. 2006. Sea lice fact sheet. http://www.aquacultureassociation .ca/news/Sea_Lice_Fact_Sheet.pdf.

AquaNet. 2005. Canada's research network in aquaculture. http://www.aquanet.ca.

ASF (Atlantic Salmon Federation). n.d. ASF policy statement on aquaculture. http://www .asf.ca.

Atkinson, C. 2007. Band goes to court to block farms. *Globe and Mail,* 28 April, S4.

Attewell, P. 1987. The deskilling controversy. *Work and Occupations* 14 (3): 323-46.

Avio, K.L. 1994. Aboriginal property rights in Canada: A contractarian interpretation of *R. v. Sparrow. Canadian Public Policy* 20 (4): 415-29.

Bael, D., and R. Sedjo. 2006. Toward globalization of the forest products industry. Resources for the Future, Discussion paper RFF DP 06-35. Washington, DC.

Bailey, C., S. Jentoft, P. Sinclair, eds. 1996. *Aquacultural development: Social dimensions of an emerging industry.* Boulder, CO: Westview Press.

Barman, J. 1996. *The west beyond the west: A history of British Columbia,* revised edition. Toronto: University of Toronto Press.

Barnes, B. 1974. *Scientific knowledge and sociological theory.* London: Routledge.

Barnett, C. 2005. The consolations of "neoliberalism." *Geoforum* 36: 7-12.

Bastien, Y. 2004. Recommendations for change. Report of the Commissioner for Aquaculture Development to the Minister of Fisheries and Oceans Canada. Fisheries and Oceans Canada, http://www.dfo-mpo.gc.ca/aquaculture.

BC Salmon Marketing Council. 2001. Salmon market database. http://www.bcsalmon.ca/dbase_local/home.html.

BC Statistics. 2002. British Columbia's fisheries and aquaculture sector. September 2002. http://www.llbc.leg.bc.ca.

BCSFA (British Columbia Salmon Farmers Association). 2005a. Sustaining members. http://www.salmonfarmers.org.

–. 2005b. Food safety. http://www.salmonfarmers.org.

–. 2005c. Organic salmon farming: Our perspective. http://www.salmonfarmers.org.

–. 2005d. Working in the BC salmon farming industry. http://www.salmonfarmers.org.

–. 2005e. *BC Salmon Farmers Association code of practice.* http://www.salmonfarmers.org.

–. 2006. Hot topics: Sea lice, wild salmon and farmed salmon. http://www.salmonfarmers.org.

–. 2007. Economic benefit and public support for aquaculture confirmed. Media release. 16 April 2007.

–. n.d. Salmon farming: The key to coastal revitalization. http://www.salmonfarmers.org.

BCSGA (British Columbia Shellfish Growers Association). 2001. Environmental management system code of practice. http://bcsga.netfirms.com.

Beamish, R., S. Jones, S. Dawe, E. Gordon, R. Sweeting, C. Neville, S. Johnson, et al. 2004. Prevalence, intensity, and life history of sea lice on adult Pacific salmon returning to the spawning areas in the Central Coast of British Columbia. Fisheries and Oceans Canada.

Beamish, R., C. Neville, R. Sweeting, and R. Ambers. 2005. Sea lice on adult Pacific salmon in the coastal wasters of British Columbia, Canada. *Fisheries Research* 76: 198-208.

Beck, U. 1992. *Risk society: Towards a new modernity.* Thousand Oaks, CA: Sage.

–. 1999. *World risk society.* Cambridge, UK: Polity Press.

Beer, A., T. Clower, G. Haughtow, and A. Maude. 2005. Neoliberalism and the institutions for regional development in Australia. *Geographical Research* 43 (1): 49-58.

Bell, A. 1991. *The language of the news media.* Malden, MA: Blackwell.

Benford, R., and D. Snow. 2000. Framing processes and social movements. *Annual Review of Sociology* 26: 611-39.

Birch, D. 1987. *Job creation in America: How our smallest companies put the most people to work.* New York: Free Press.

Bivens, J. 2003. Updated employment multipliers for the U.S. economy. Working paper no. 268. Washington, DC: Economic Policy Institute.

Blaikie, N. 2003. *Analyzing quantitative data.* Thousand Oaks, CA: Sage.

Blanchflower, D., and A. Oswald. 1999. Well-being, insecurity, and the decline of American job security. Dartmouth University, http://www.dartmouth.edu/~blnchflr/papers/JobSat.pdf.

Blok, A. 2007. Experts on public trial: On democratizing expertise through a Danish consensus conference. *Public Understanding of Science* 16: 163-82.

Bloor, D. 1976. *Knowledge and social imagery.* London: Routledge.

Blumenthal, P., and J.L. Jannik. 2000. A classification of collaborative management methods. *Conservation Ecology* 4 (2): 13.

Bomsel, O., P. Borkey, M. Glachant, and F. Levesque. 1996. Is there room for environmental self-regulation in the mining sector? *Resources Policy* 22 (1-2): 79-86.

Bonnen, J. 2000. The transformation of agriculture and the world economy. In *Tomorrow's agriculture,* ed. G.H. Peters and P. Pingali, 12-37. Oxford: Ashgate.

Boykoff, M., and J. Boykoff. 2004. Balance as bias: Global warming and the US prestige press. *Global Environmental Change* 14: 125-36.

Braverman, H. 1974. *Labor and monopoly capitalism: The degradation of work in the twentieth century.* New York: Monthly Review Press.

Brenner, N. 2003. "Glocalization" as a state spatial strategy: Urban entrepreneurialism and the new politics of uneven development in western Europe. In *Remaking the global economy: Economic-geographical perspectives,* ed. J. Peck and H. Yeung, 197-215. Thousand Oaks, CA: Sage.

–. 2004. *New state spaces: Urban governance and the rescaling of statehood.* Oxford: Oxford University Press.

Brenner, N., and N. Theodore. 2002. Cities and the geographies of "actually existing neo-liberalisms." *Antipode* 34 (3): 349-79.

Brenner, N., and N. Theodore. 2005. Neoliberalism and the urban condition. *City* 9: 101-7.

British Columbia. 2002a. Funding of scientific research on aquaculture. Media release, Ministry of Agriculture, Food and Fisheries, 12 September 2002.

–. 2002b. BC shellfish aquaculture code of practice (Ministry of Agriculture, Food and Fisheries). The David Suzuki Foundation, http://www.davidsuzuki.org/files/Oceans/FinalCOPSubmission02July03.pdf.

–. 2002c. Finfish aquaculture waste control regulation. Media backgrounder, Ministry of Agriculture, Food and Fisheries.

–. 2002d. Quarterly progress report. Minister of State for Deregulation, http://www.llbc.leg.bc.ca/public/pubdocs/bcdocs/352458/Sept_2002/Sept2002.pdf.

–. 2003a. The forestry revitalization plan. Ministry of Forests, http://www.for.gov.bc.ca/mof/plan/frp.

–. 2003b. Backgrounder: Broughton Archipelago action plan. 6 February 2003. Ministry of Agriculture, Food and Fisheries, http://www.llbc.leg.bc.ca/public/PubDocs/bcdocs/359361/backgrounder.pdf

–. 2006a. Minutes of the Special Committee on Sustainable Aquaculture. 2005 Legislative Session: First Session, 38th Parliament of the Legislature of British Columbia. Legislative Assembly of British Columbia, http://www.leg.bc.ca/CMT/38thparl/session-1/aquaculture/index.htm.

–. 2006b. Strong fish health strategy working, audit shows. Media release 2006AL0046-001466, Ministry of Agriculture and Lands.

–. 2007a. Shellfish development initiative. Ministry of Agriculture and Lands, http://www.agf.gov.bc.ca/fisheries.

–. 2007b. Tenure fee reduction to boost shellfish opportunities. Media release, 9 June. Ministry of Agriculture and Lands.

–. 2008a. Seafood data tables and graphs. Ministry of the Environment, http://www.env.gov.bc.ca/omfd/fishstats/graphs-tables/index.html.

–. 2008b. Finfish aquaculture suspended in the North Coast area. Media release, 27 March. Ministry of Agriculture and Lands.

Brooks, K. 2006. A critical review of Krkosek et al. BC Salmon Farmers Association, http://www.salmonfarmers.org.

Brown, D. 2005. *Salmon wars: The battle for the west coast salmon fishery.* Madeira Park, BC: Harbour Publishing.

Brundtland, G.H. 1987. *Our common future: Report of the World Commission on Environment and Development.* New York: Oxford University Press.

Bryden, J., and R. Bollman. 2000. Rural employment in industrialized countries. *Agricultural Economics* 22: 185-97.

Bryson, B. 2003. *A short history of nearly everything.* Toronto: Anchor.

Burda, C., and F. Gale. 1998. Trading in the future: British Columbia's forest products compromise. *Society and Natural Resources* 11 (6): 555-69.

Burns, T., D. O'Connor, and S. Stocklmayer. 2003. Science communication: A contemporary definition. *Public Understanding of Science* 12: 183-202.

Butterworth, K., F. Cubitt, B. Finstad, and S. McKinley. 2006. Sea lice: The science behind the hype. Fraser Institute, http://www.fraserinstitute.org.

CAAR (Coastal Alliance for Aquaculture Reform). n.d.a. Flyer handout. http://www.farmedanddangerous.org.

–. n.d.b. Salmon farming and First Nations. http://www.farmedanddangerous.org.

–. 2005. BC organic associations rule farmed salmon raised in netcages ineligible for certification. Media release, 1 March.

Cadigan, S. 2006. Restructuring in the woods: Timber rights, power, and agency in White Bay, Newfoundland, 1897-1959. In *Power and restructuring: Canada's coastal society and environment,* ed. R. Ommer and P. Sinclair, 54-81. St. John's: ISER Books.

CAIA (Canadian Aquaculture Industry Alliance). 2005a. Markets and statistics. http://www.aquaculture.ca.

–. 2005b. Environmental scan. http://www.aquaculture.ca.

–. 2006a. Canadian aquaculture industry profile. http://www.aquaculture.ca.

–. 2006b. Sea lice study based on old, questionable data. News release, 4 October.

Cairns, A. 2000. *Citizens plus: Aboriginal peoples and the Canadian state.* Vancouver: UBC Press.

Canada. 2001. Departmental performance report, 2000-2001. Agriculture and Agri-Food Canada; Treasury Board of Canada. http://www.tbs-sct.gc.ca/tbs-sct/cmn/archives-eng.asp.

–. 2006. The Canadian farmed salmon industry: Benchmark analysis for the US market. Agriculture and Agri-Food Canada, http://www.ats.agr.gc.ca.

Carroll, W., and R. Ratner. 1996. Master frames as counter-hegemony: Political sensibilities in contemporary social movements. *Canadian Review of Sociology and Anthropology* 33: 407-35.

Carson, R. 1962. *Silent Spring.* Boston: Houghton Mifflin.

Carsten, A. 2002. Commentary on the draft code of practice for the shellfish industry. Alliance for Responsible Shellfish Farming, http://www.responsibleshellfishfarming.ca.

Cartmel, F., and A. Furlong. 2000. *Youth unemployment in rural areas.* York, UK: York Publishing Services.

Carvalho, A. 2007. Ideological cultures and media discourses on scientific knowledge: Rereading the news on climate change. *Public Understanding of Science* 16: 223-43.

Castells, M. 2000. *The Rise of the network society,* 2nd ed. Cambridge, MA: Blackwell.

Castree, N. 2006. Commentary. *Environment and Planning A* 38: 1-6.

Cayo, D. 1993. Silver harvest: New Brunswick's salmon farms are reviving a rural economy. *Canadian Geographic* (July/August).

CCFAM (Canadian Council of Fisheries and Aquaculture Ministers). 2001. Canadian action plan for aquaculture. Fisheries and Oceans Canada, http://www.dfo-mpo.gc.ca/Aquaculture.

Che, A. 2005. Remote biology labs. Paper presented at the MIT Conference on Education without Borders, 19-21 February.

Chowdhury, A., and M. Jakariya. 1999. Testing of water for arsenic in Bangladesh. *Science* 284 (5420): 1621.

Chryssochoidis, G., A. Strada, and A. Krystallis. 2009. Public trust in institutions and information sources regarding risk management: Towards integrating extant knowledge. *Journal of Risk Research* 12 (2): 137-85.

Coates, K. 2000. *The Marshall decision and Native rights.* Montreal: McGill-Queen's University Press.

Coglianese, C. 2003. Designing and implementing performance-based regulation. Presentation to the Government of British Columbia. http://www.dcat.net/resources/ircc_global_policy_summit-final_report.pdf.

Cole, H., ed. 1973. *Models of doom: A critique of the limits to growth.* New York: Universe Books.

Collins, H. 1974. The TEA-set: Tacit knowledge and scientific networks. *Science Studies* 4: 165-86.

–. 1983. The sociology of scientific knowledge: Studies of contemporary science. *Annual Review of Sociology* 9: 265-83.

Collins, H., and R. Evans. 2002. The third wave of science studies: Studies of expertise and experience. *Social Studies of Science* 32 (2): 235-96.

Connell, I. 1998. Mistaken identities: Tabloid and broadsheet news discourses. *The Public* 5 (3): 11-31.

Covello, V., M. Pavlova, and D. McCallum. 1989. *Effective risk communication.* New York: Plenum.

Cox, R. 2006. *Environmental communication and the public sphere.* Thousand Oaks, CA: Sage.

Cox, S. 2004. *Diminishing returns: An investigation into the five multinational corporations that control British Columbia's salmon farming industry.* Vancouver: Raincoast Conservation Society; Coastal Alliance for Aquaculture Reform.

Cronbach, L. 1951. Coefficient alpha and the internal structure of tests. *Psychometrika* 16: 297-334.

Cronon, W. 1995. The trouble with wilderness. In *Uncommon ground,* ed. W. Cronon, 69-91. New York: Norton.

Cvetkovitch, G. 1999. The attribution of social trust. In *Social trust and the management of risk,* ed. G. Cvetkovitch and R. Lofstedt, 53-61. London: Earthscan.

Dacks, G. 2002. British Columbia after the *Delgamuukw* decision: Land claims and other processes. *Canadian Public Policy* 28 (2): 239-55.

Deal, H. 2005. *Sustainable shellfish: Recommendations for responsible aquaculture.* Vancouver: David Suzuki Foundation.

Dean, M. 1999. *Governmentality.* Thousand Oaks, CA: Sage.

Dearing, J. 1995. Newspaper coverage of maverick science: Creating controversy through balancing. *Public Understanding of Science* 4: 341-61.

Demeritt, D. 2000. The new social contract for science: Accountability, relevance, and value in US and UK science and research policy. *Antipode* 32 (3): 308-29.

DFO (Department of Fisheries and Oceans). 2000. DFO's aquaculture action plan. http://www.dfo-mpo.gc.ca/aquaculture.

–. 2002. DFO's aquaculture policy framework. http://www.dfo-mpo.gc.ca/aquaculture.

–. 2004. A policy framework for the management of fisheries on Canada's Atlantic coast. http://www.dfo-mpo.gc.ca/fm-gp/policies-politiques/afpr-rppa/framework-cadre-eng.pdf.

–. 2005a. Canadian aquaculture industry, 2004-2005: Key figures. http://www.dfo-mpo.gc.ca/aquaculture.

–. 2005b. Aquaculture in Eastern Canada – A growing opportunity. http://www.dfo-mpo.gc.ca/aquaculture.

–. 2006a. DFO's aquaculture action plan: Interim guide to fisheries resource use considerations in the evaluation of aquaculture site applications. http://www.dfo-mpo.gc.ca/aquaculture.

–. 2006b. DFO's aquaculture action plan: Interim guide to information requirements for environmental assessment of marine finfish aquaculture projects. http://www.dfo-mpo.gc.ca/aquaculture.

–. 2006c. DFO's aquaculture action plan: Interim guide to the application of section 35 of the Fisheries Act to salmonid cage aquaculture developments. http://www.dfo-mpo.gc.ca/aquaculture.

–. 2006d. State of knowledge presentation for the Special Committee on Sustainable Aquaculture of the British Columbia legislature. http://www.pac.dfo-mpo.gc.ca/science/aquaculture/sok-edc/exec-eng.htm.

–. 2006e. 2000-2004 integrated fisheries management plan – oyster Prince Edward Island. http://www.glf.dfo-mpo.gc.ca/fam-gpa/plans/pei-ipe/oysters_huitres_2000_2004-e.php.

–. 2007. Transcript: Farming the seas. http://www.dfo-mpo.gc.ca/aquaculture/video-eng.htm.

–. 2008. Aquaculture collaborative research and development program – program information. http://www.dfo-mpo.gc.ca/science/aquaculture/acrdp_e.htm.

Dhaliwal, H. 2000. Speech of the Minister of Fisheries and Oceans to announce funding for aquaculture development in Canada. Given at Vancouver, BC, 8 August. Fisheries and Oceans Canada, http://www.dfo-mpo.gc.ca.

Doern, G., and T. Reed. 2000. Canada's changing science-based policy and regulatory regime: Issues and framework. In *Risky business: Canada's changing science-based policy and regulatory regime,* ed. B. Doern and T. Reed, 3-28. Toronto: University of Toronto Press.

Douglas, M. 1966. *Purity and danger: An analysis of the concepts of pollution and taboo.* New York: Routledge.

Douglas, M., and A. Wildavsky. 1982. *Risk and culture.* Berkeley: University of California Press.

Downs, A. 1972. Up and down with ecology: The issue attention cycle. *Public Interest* (Summer): 38-50.

Dryzek, J. 2005. *The politics of the earth.* New York: Oxford University Press.

DSF (The David Suzuki Foundation). 2007. Ad campaigns: Salmon aquaculture. http://www.davidsuzuki.org/Ad_Campaigns/Executive_Summary.asp.

DSF-RCS (The David Suzuki Foundation; Raincoast Conservation Society). 2005. Fish farm link to sea lice infections on BC wild salmon confirmed. Science bulletin. Media release.

Dunwoody, S. 1999. Scientists, journalists, and the meaning of uncertainty. In *Communicating uncertainty: Media coverage of new and controversial science,* ed. S. Friedman, S. Dunwoody, and C. Rogers, 59-79. Mahwah, NJ: Lawrence Ehrlbaum Associates.

Durant, R., D. Fiorino, and R. O'Leary, eds. 2004. *Environmental governance reconsidered: Challenges, choices, opportunities.* Cambridge, MA: MIT Press.

Dutrisac, R. 2008. Québec veut libérer la forêt. *Le Devoir,* 15 February. http://www.ledevoir.com.

EACSR. 2004. *Smart regulation: A regulatory strategy for Canada.* Ottawa: Government of Canada.

Easton, M., and D. Luszniak. 2002. Preliminary examination of contaminant loadings in farmed salmon, wild salmon and commercial salmon feed. *Environmental Science and Technology* 40 (11): 3489-93.

Economist. 2003. The promise of a blue revolution: How aquaculture might meet most of the world's demand for fish without ruining the environment. *Economist,* 7 August, 20-3.

Ecotrust Canada. 2004. *Catch-22: Conservation, communities, and the privatization of BC fisheries.* Vancouver: Ecotrust Canada.

Edgar, P. 2002. More action pledged after fish farm attack. *Vancouver Sun,* 20 December, B2.

Ehrenfeld, D. 2005. The environmental limits of globalization. *Conservation Biology* 19 (2): 318-26.

Einsiedel, E., and E. Coughlan. 1993. The Canadian press and the environment: Reconstructing a social reality. In *The mass media and environmental issues,* ed. A. Hansen, 134-49. New York: Leicester University Press.

Ellis, D. 1996. *Net loss: The salmon netcage industry in British Columbia.* Vancouver: David Suzuki Foundation.

Environmental Law Centre. 2006. *Shellfish aquaculture and the new relationship.* 17 November 2006. Victoria: University of Victoria. http://www.elc.uvic.ca/documents/shellfish-aquaculture-background-paper.pdf.

Environmental Working Group. 2003. Factory methods, unnatural results. http://www.ewg.org/reports/farmedPCBs.

Epp, R., and D. Whitson, eds. 2001. *Writing off the rural west: Globalization, governments, and the transformation of rural communities.* Edmonton: University of Alberta Press.

Ericson, R., and A. Doyle, eds. 2003. *Risk and morality.* Toronto: University of Toronto Press.

FAO (Food and Agriculture Organization). 2004. *The state of world fisheries and aquaculture, 2004.* Rome: United Nations.

–. 2006. *The state of world fisheries and aquaculture, 2006.* Rome: United Nations.

–. 2007a. FAO Forestry Statistics Service. http://www.fao.org/forestry.

–. 2007b. FAO Fisheries Statistics Service. http://www.fao.org/fishery/statistics/programme.

–. 2008. *The state of world fisheries and aquaculture, 2008.* Rome: United Nations.

Farm Credit Canada. 2004. Farm Credit Canada provides $3.4 billion to agriculture. Media release, 8 October.

FFAW (Fish, Food and Allied Workers Union). 2001. Response to Atlantic fisheries policy review. http://www.ffaw.nf.ca/IssuesCampaigns.asp.

Fiorino, D. 1990. Citizen participation and environmental risk: A survey of institutional mechanisms. *Science, Technology and Human Values* 15 (2): 226-43.

Fischer, F. 2000. *Citizens, experts, and the environment.* Durham, NC: Duke University Press.

Fischhoff, B. 1995. Risk perception and communication unplugged: Twenty years of process. *Risk Analysis* 15 (2): 137-45.

Fish Farmer. 2006. Chile: Salmon fillet export revenue up 27%. 16 August. http://www.fishfarmer-magazine.com

Forsyth, T. 2003. *Critical political ecology: The politics of environmental science.* New York: Routledge.

Fox, G. 2006. Mediating resource management in the Mi'kmaq fisheries. *Development* 49: 119-24.

FRCC. 2007. *Sustainability framework for Atlantic lobster, 2007.* Ottawa: Ministry of Public Works and Government Services.

Friedman, S., S. Dunwoody, and C. Rogers, eds. 1999. *Communicating uncertainty: Media coverage of new and controversial science.* London: Lawrence Erlbaum Associates.

Gallagher, R., and T. Appenzeller. 1999. Beyond reductionism. *Science* 284 (5411): 79.

Gamson, W., and A. Modigliani. 1989. Media discourse and public opinion on nuclear power. *American Journal of Sociology* 95 (1): 1-37.

Gardner, J., and D. Peterson. 2003. *Making sense of the salmon aquaculture debate.* Report to the Pacific Fisheries Resource Conservation Council. Vancouver.

Georgia Strait Alliance. 2002. Environmentalists quit salmon farming advisory body. Media release, 1 February.

–. 2004. *Regulating salmon aquaculture in BC: A report card.* Prepared for the Coastal Alliance for Aquaculture Reform, Nanaimo, BC.

Giddens, A. 1990. *The consequences of modernity.* Stanford, CA: Stanford University Press.

–. 2000. *Runaway world.* New York: Routledge.

Gillespie, D. 1986. *An inquiry into finfish aquaculture in British Columbia: Report and recommendations.* Victoria: Government of British Columbia.

Ginetz, R. 2002. *On the risk of colonization by Atlantic salmon in BC waters.* Campbell River, BC: BC Salmon Farmers Association.

Gislason and Associates. 2006. *The Canadian farmed salmon industry: Benchmark analysis for the US market.* Prepared for Agriculture and Agri-Food Canada.

Gleditsch, N., ed. 1997. *Conflict and the environment.* Norwell, MA: Kluwer Academic.

Goldin, I. 1990. *Comparative advantage: Theory and application to developing country agriculture.* Paris: Organisation for Economic Co-operation and Development.

Goodstein, E. 1999. *The trade-off myth: Fact and fiction about jobs and the environment.* Washington, DC: Island Press.

Grafton, Q., and D. Lane. 1998. Canadian fisheries policy: Challenges and choices. *Canadian Public Policy* 24 (2): 133-47.

Graham, S., and S. Marvin. 2001. *Splintering urbanism: Networked infrastructures, technological mobilities, and the urban condition.* New York: Routledge.

Gramsci, A. 1992 [1935]. *Prison notebooks.* New York: Columbia University Press.

Gray, A. 2004. Ecology and government policies: The GM crop debate. *Journal of Applied Ecology* 41 (1): 1-10.

Gray, G., and N. Guppy. 1999. *Successful surveys: Research methods and practice.* Toronto: Harcourt Brace.

Greenpeace. 2004. Keep it wild! No fish farms. http://www.greenpeace.org/usa/news/keep-it-wild-no-fish-farms.

Grieder, T., R. Krannich, and H. Berry. 1991. Local identity, solidarity, and trust in changing rural communities. *Sociological Focus* 24: 263-82.

Guber, D. 2003. *The grassroots of a green revolution: Polling America on the environment.* Cambridge, MA: MIT Press.

Gunningham, N., and J. Rees. 1997. Industry self-regulation: An institutional perspective. *Law and Policy* 19: 363-414.

Gunningham, N., P. Grabosky, and D. Sinclair. 1998. *Smart regulation: Designing environmental policy.* Oxford: Clarendon Press.

Gunningham, N., R. Kagan, and D. Thornton. 2004. Social licenses and environmental protection: Why businesses go beyond compliance. *Law and Social Inquiry* 29: 307-42.

Gunther, A., and K. Schmitt. 2004. Mapping the boundaries of the hostile media effect. *Journal of Communication* 54 (1): 55-70.

Hacking, I. 1999. *The social construction of what?* Cambridge, MA: Harvard University Press.

Hall, S., C. Critcher, T. Jefferson, J. Clarke, and B. Roberts. 1978. *Policing the crises: Mugging, the state, and law and order.* New York: Holmes and Meier.

Halse, N. n.d. Fishin' for a new deal: Why aquaculture belongs with agriculture. Press release. Canadian Aquaculture Industry Alliance, http://www.aquaculture.ca/media/PressReleases/ CAIA_PressReleases91.htm.

Halseth, G., and R. Halseth, eds. 2004. *Building for success: Explorations of rural community and rural development.* Brandon, MB: Canadian Rural Revitalization Foundation.

Hannigan, J. 2006. *Environmental sociology.* New York: Routledge.

Hansen, A., ed. 1993. *The mass media and environmental issues.* New York: Leicester University Press.

Haraway, D. 1988. Situated knowledges: The science question in feminism and the privilege of partial perspective. *Feminist Studies* 14 (3): 575-99.

Harding, S. 1991. *Whose science? Whose knowledge? Thinking from women's lives.* Ithaca, NY: Cornell University Press.

Hardwig, J. 1991. The role of trust in knowledge. *Journal of Philosophy* 88 (12): 693-708.

Harremoes, P., ed. 2002. *The precautionary principle in the 20th century.* London: Earthscan.

Harris, Cole. 2002. *Making native space: Colonialism, resistance, and reserves in British Columbia.* Vancouver: UBC Press.

Harrison, K. 2001. Voluntarism and environmental governance. In *Governing the environment: Persistent challenges, uncertain innovations,* ed. E. Parson, 207-46. Toronto: University of Toronto Press.

Hartz, J., and C. Chappell. 1997. *Worlds apart: How the distance between science and journalism threatens America's future.* Nashville, TN: First Amendment Center.

Harvey, D. 1989. From managerialism to entrepreneurialism: The transformation in urban governance in late capitalism. *Geografiska Annaler Series B* 71: 3-17.

Harvey, D.J. 2006. Aquaculture outlook. Document no. LDP-AQS-24. Washington, DC: US Department of Agriculture.

Hayter, R. 2000. *Flexible crossroads.* Vancouver: UBC Press.

–. 2003. The war in the woods: Post-Fordist restructuring, globalization, and the contested remapping of British Columbia's forest economy. *Annals of the Association of American Geographers* 93 (3): 706-29.

Hayter, R., and T. Barnes. 1997. The restructuring of British Columbia's coastal forest sector: Flexibility perspectives. *BC Studies* 113: 7-34.

Hegele, R., F. Sun, S. Harris, C. Anderson, A. Hanley, and B. Zinman. 1999. Genome-wide scanning for type 2 diabetes susceptibility in Canadian Oji-Cree. *Journal of Human Genetics* 44: 10-14.

Herbert-Cheshire, L. 2003. Translating policy: Power and action in Australia's country towns. *Sociologia Ruralis* 43 (4): 454-73.

Herbert-Cheshire, L., and V. Higgins. 2004. From risky to responsible: Expert knowledge and the governing of community-led development. *Journal of Rural Studies* 20: 289-302.

Hicks, B. 2004. Letter to *Science* magazine editor, 27 February. BC Salmon Farmers Association, http://www.salmonfarmers.org.

Hindar, K. 2003. Wild Atlantic salmon in Europe: Status and perspectives. In *Proceedings from the World Summit on Salmon.* Burnaby: Simon Fraser University. http://www.sfu.ca/ cstudies/science/resources/summit/pdf/05%20-%20Hindar.pdf.

Hirst, P., and G. Thompson. 1996. *Globalization in question.* Cambridge, UK: Polity Press.

Hites, R., J. Foran, D. Carpenter, C. Hamilton, B. Knuth, and S. Schwager. 2004. Global assessment of organic contaminants in farmed salmon. *Science* 303 (9 January): 226-29.

Hoijer, B., R. Lidskog, and L. Thornberg. 2006. News media and food scares: The case of contaminated salmon. *Environmental Sciences* 3 (4): 273-88.

Horne, G. 2004. *British Columbia's heartland at the dawn of the 21st century: 2001 economic dependencies and impact ratios for 63 local areas.* Victoria: BC Ministry of Management Services.

Howlett, M. 2000. Managing the hollow state: Procedural policy instruments and modern governance. *Canadian Public Administration* 43 (4): 412-31.

–. 2001. Complex network management and the governance of the environment: Prospects for policy change and policy stability over the long term. In *Governing the environment: Persistent challenges, uncertain innovations,* ed. E. Parson, 303-44. Toronto: University of Toronto Press.

–. 2003. Canadian environmental policy and the natural resources sector: Paradoxical aspects of the transition to a post-staples political economy. In *The integrity gap: Canada's environmental policy and institutions,* ed. E. Lee and A. Pearl, 42-67. Vancouver: UBC Press.

Howlett, M., and K. Brownsey, eds. 2008. *Canada's resource economy in transition: The past, present, and future of Canadian stapes industries.* Toronto: Emond Montgomery.

Howlett, M., and J. Rayner. 2004. (Not so) smart regulation? Canadian shellfish aquaculture policy and the evolution of instrument choice for industrial development. *Marine Policy* 28 (2): 171-84.

Hume, S., A. Morton, B. Keller, R. Leslie, O. Langer, and D. Staniford. 2004. *A stain upon the sea: West coast salmon farming.* Madeira Park, BC: Harbour Publishing.

INAC. 2003. Milbrook First Nation: Aquaculture with the future in mind. *Aboriginal Economic Development in Atlantic Canada* 2003: 3.

–. 2005. Aqua-Culture Nordik Fish Farm opens for business in Port-Daniel Gascons. Media release, 9 August.

Ignatieff, M. 2000. *The rights revolution.* Toronto: House of Anansi Press.

Industry Canada. 2008. Canadian trade by industry. Trade data online. http://www.ic.gc.ca/eic/site/tdo-dcd.nsf/eng/Home.

Innis, H. 1933. *Problems of staple production in Canada.* Toronto: Ryerson Press.

–. 1956. *Essays in Canadian economic history.* Toronto: University of Toronto Press.

Irwin, A. 1995. *Citizen science.* New York: Routledge.

–. 2001. *Sociology and the environment.* Cambridge, UK: Polity Press.

Irwin, A., and B. Wynne, eds. 1996. *Misunderstanding science?* New York: Cambridge University Press.

Iyengar, S., and A. Simon. 1993. News coverage of the Gulf crisis and public opinion. *Communication Research* 20 (3): 365-83.

Jackson, L., E.A. Marshall, S. Tirone, C. Donovan, and B. Shepard. 2006. The forgotten population? Power, powerlessness, and agency among youth in coastal communities. In *Power and restructuring: Canada's coastal society and environment,* ed. P. Sinclair and R. Ommer, 232-49. St. John's: ISER Books.

Jackson, T., and J. Curry. 2004. Peace in the woods: Sustainability and the democratization of land use planning and resource management on Crown lands in British Columbia. *International Planning Studies* 9 (1): 27-42.

Jacobs, M., N. Covaci, and P. Schepens. 2002a. Investigation of selected persistent organic pollutants in farmed Atlantic salmon, salmon aquaculture feed, and fish oil components of the feed. *Environmental Science and Technology* 36 (13): 2792-2805.

Jacobs, M., J. Ferrario, and C. Byrne. 2002b. Investigation of polychlorinated dibenzo-*p*-dioxins in Scottish farmed Atlantic salmon. *Chemosphere* 47 (2): 183-91.

Jasanoff, S. 2004. Heaven and earth: The politics of environmental images. In *Earthly Politics,* ed. S. Jasanoff and M. Martello, 31-52. Cambridge, MA: MIT Press.

Jessop, B. 2002a. *The future of the capitalist state.* Malden, MA: Polity Press.

–. 2002b. Liberalism, neoliberalism, and urban governance. *Antipode* 34: 452-72.

–. 2004. From localities via the spatial turn to spatio-temporal fixes: A strategic-relational odyssey. Discussion forum, Socio-Economics of Space Group at University of Bonn, http://www.giub.uni-bonn.de/grabher/downloads/Jessop.pdf.

Joffe, P. 2000. Assessing the *Delgamuukw* principles: National implications and potential effects in Quebec. *McGill Law Journal* 45: 155-208.

Johnson, H. 1971. The Keynesian revolution and the monetarist counter-revolution. *American Economic Review* 61 (2): 1-14.

Joyce, S. 2000. Earning a social licence to operate: Social acceptability and resource development in Latin America. *CIM Magazine* 93: 1037.

Kallerud, E., and I. Ramberg. 2002. The order of discourse in surveys of public understanding of science. *Public Understanding of Science* 11: 213-24.

Kaplowitz, M., T. Hadlock, and R. Levine. 2004. A comparison of web and mail survey response rates. *Public Opinion Quarterly* 68 (1): 94-101.

Kasperson, R., O. Renn, P. Slovic, H. Brown, J. Emel, R. Goble, J. Kasperson, et al. 1988. The social amplification of risk: A conceptual framework. *Risk Analysis* 8 (2): 177-87.

Kasperson, R., J. Nayna, and J. Kasperson. 2001. Stigma and the social amplification of risk. In *Risk, media and stigma: Understanding public challenges to modern science and technology,* ed. James Flynn, Paul Slovic, and Howard Kunreuther, 9-30. London: Earthscan.

Keller, B., and R. Leslie. 1996. *Sea-silver: Inside British Columbia's salmon farming industry.* Victoria: Horsdal and Schubart.

Kennedy, J.C. 2006. Disempowerment: The cod moratorium, fisheries restructuring, and the decline of power among Labrador fishers. In *Power and restructuring,* ed. P. Sinclair and R. Ommer, 129-44. St. John's: ISER Books.

Kenney, A. 1997. *Net gain: The salmon farming industry in British Columbia.* Prepared for the British Columbia Salmon Farmers Association, Vancouver.

Kerr, A., S. Cunningham-Burley, and R. Tutton. 2007. Shifting subject positions: Experts and lay people in public dialogue. *Social Studies of Science* 37 (3): 385-411.

Killingsworth, J., and J. Palmer. 1992. *Ecospeak: Rhetoric and environmental politics in America.* Carbondale, IL: Southern Illinois University Press.

Kinder, D., and L. Sanders. 1990. Mimicking political debate with survey questions: The case of white opinion on affirmative action for blacks. *Social Cognition* 8 (1): 73-103.

King, L., and D. McCarthy, eds. 2005. *Environmental sociology.* Lanham, MD: Rowman and Littlefield.

Klein, K., and W. Kerr. 1995. The globalization of agriculture: A view from the farm gate. *Canadian Journal of Agricultural Economics* 43 (4): 551-63.

Knapp, G., C. Roheim, and J. Anderson. 2007. *The great salmon run: Competition between wild and farmed salmon.* Washington, DC: World Wildlife Fund.

Knoke, D., and G. Bohrnstedt. 1991. *Basic social statistics.* Itasca, IL: F.E. Peacock.

Knorr Cetina, K. 1982. The ethnographic study of scientific work. In *Science observed,* ed. K. Knorr Cetina and M. Mulkay, 115-40. London: Sage.

–. 1999. *Epistemic cultures: How the sciences make knowledge.* Cambridge, MA: Harvard University Press.

Krkosek, M. 2005. Fish farm link to sea lice infections on BC wild salmon confirmed. Pamphlet. Vancouver: David Suzuki Foundation and Raincoast Conservation Society.

Krkosek, M., M. Lewis, and J. Volpe. 2005a. Transmission dynamics of parasitic sea lice from farm to wild salmon. *Proceedings of the Royal Society B* (30 March): 689-96.

Krkosek, M., A. Morton, and J. Volpe. 2005b. Non-lethal assessment of juvenile pink salmon and chum salmon for parasitic sea lice infections and fish health. *Transactions of the American Fisheries Society* 134: 711-16.

Krkosek, M., M. Lewis, A. Morton, N. Frazer, and J. Volpe. 2006. Epizootics of wild fish induced by farm fish. *Proceedings of the National Academy of Sciences of the United States of America* 103 (42): 15506-10.

Kuhn, T. 1970. *The structure of scientific revolutions.* Chicago: University of Chicago Press.

–. 1977. *The essential tension: Selected studies in scientific tradition and change.* Chicago: University of Chicago Press.

Langer, O. 2004. Any fish is a good fish: Salmon farming and the Department of Fisheries and Oceans. In *A stain upon the sea: West coast salmon farming,* ed. S. Hume et al., 123-44. Madeira Park, BC: Harbour Publishing.

Larner, W. 2003. Neoliberalism? *Environment and Planning D* 21: 509-12.

Lash, S., and J. Urry. 1987. *The end of organized capitalism.* Cambridge, UK: Polity Press.

Latour, B., and S. Woolgar. 1979. *Laboratory life: The social construction of scientific facts.* Beverly Hills, CA: Sage.

Lawton, J. 1999. Are there general laws in ecology? *Oikos* 84 (2): 177-92.

Leggatt, S. 2002. *Clear choices, clean waters.* David Suzuki Foundation, http://www.davidsuzuki.org/files/Leggatt_reportfinal.pdf.

Leiss, W. 2000. Between expertise and bureaucracy: Risk management trapped at the science/policy interface. In *Risky business,* ed. G.B. Doern and T. Reed, 50-74. Toronto: University of Toronto Press.

–. 2001. *In the chamber of risk.* Montreal and Kingston: McGill-Queen's University Press.

Leiss, W., and C. Chociolko. 1994. *Risk and responsibility.* Montreal and Kingston: McGill-Queen's University Press.

Leiss, W., and A. Nicol. 2006. A tale of two food risks: BSE and farmed salmon in Canada. *Journal of Risk Research* 9 (8): 891-910.

Levi-Faur, D. 2005. The global diffusion of regulatory capitalism. *Annals of the American Academy of Political and Social Science* 598: 12-32.

Levins, R., and R. Lewontin. 1980. Dialectics and reductionism in ecology. *Synthese* 43: 47-78.

Levy-Leblond, J. 1992. Misunderstandings about misunderstandings. *Public Understanding of Science* 1: 17-21.

Lewis, J.D., and A. Weigert. 1985. Trust as a social reality. *Social Forces* 63 (4): 967-85.

Lipietz, A. 1987. *Mirages and miracles: The crisis in global Fordism.* London: Verso.

Loftstedt, R., and O. Renn. 1997. The *Brent Spar* controversy: An example of risk communication gone wrong. *Risk Analysis* 17 (2): 131-36.

Longino, H. 1990. *Science as social knowledge: Values and objectivity in scientific inquiry.* Princeton, NJ: Princeton University Press.

Luhmann, N. 1979. *Trust and power.* New York: John Wiley and Sons.

Lupton, D. 1999. *Risk.* New York: Routledge.

MacMillan, H., A. MacMillan, D. Offord, and J. Dingle. 1996. Aboriginal health. *Canadian Medical Association Journal* 155 (11): 1569-78.

Macnaghten, P., and J. Urry. 1998. *Contested natures.* London: Sage.

Magnusson, W., and K. Shaw. 2002. *A political space: Reading the global through Clayoquot Sound.* Montreal and Kingston: McGill-Queen's University Press.

Malone, R., E. Boyd, and L. Bero. 2000. Science in the news: Journalists' constructions of passive smoking as a social problem. *Social Studies of Science* 30 (5): 713-35.

Mansfield, B. 2004. Organic views of nature: The debate over organic certification for aquatic animals. *Sociologia Ruralis* 44 (2): 216-32.

Marchak, P. 1983. *Green gold.* Vancouver: UBC Press.

Marchak, P., S. Aycock, and D. Herbert. 1999. *Falldown: Forest policy in British Columbia.* Vancouver: David Suzuki Foundation.

Markey, S., J. Pierce, K. Vodden, and M. Roseland. 2005. *Second growth: Community economic development in rural British Columbia.* Vancouver: UBC Press.

Marshall, D. 2003. *Fishy business: The economics of salmon farming in British Columbia.* Vancouver: Canadian Centre for Policy Alternatives.

Marshall, J. 2001. Landlords, leaseholder and sweat equity: Changing property regimes in aquaculture. *Marine Policy* 25: 335-52.

Martin, P. 1997. "If you don't know how to fix it, please stop breaking it!" The precautionary principle and climate change. *Foundations of Science* 2: 263-92.

Martin, P., S. Browne-Clayton, and E. McWilliams. 2002. A results-based system for regulating reforestation obligations. *Forestry Chronicle* 78 (4): 492-98.

Marx, K. 1985 [1848]. *The communist manifesto.* London: Penguin.

Matthews, R. 1983. *The creation of regional dependency.* Toronto: University of Toronto Press.

–. 1993. *Controlling common property.* Toronto: University of Toronto Press.

–. 2004. *The Canadian aquaculture employment study.* Report for the Canadian Aquaculture Industry Alliance. Vancouver.

Matthews, R., and N. Young. 2005. Development on the margin: Development orthodoxy and the success of Lax Kw'alaams, British Columbia. *Journal of Aboriginal Economic Development* 4 (2): 100-8.

Matthews, R., R. Pendakur, and N. Young. 2009. Social capital, labour markets, and job-finding in rural and urban regions: Comparing paths to employment in prosperous cities and stressed rural communities in Canada. *Sociological Review* 57 (2): 306-30.

Mazur, A., and J. Lee. 1993. Sounding the global alarm: Environmental issues in the US national news media. *Social Studies of Science* 23 (4): 681-720.

Mazzotta, M., and J. Kline. 1995. Environmental philosophy and the concept of non-use value. *Land Economics* 71 (2): 244-49.

McCarthy, J. 2005. Devolution in the woods: Community forestry as a hybrid neoliberalism. *Environment and Planning A* 37: 995-1014.

–. 2006. Neoliberalism and the politics of alternatives: Community forestry in British Columbia and the United States. *Annals of the Association of American Geographers* 96 (1): 84-104.

McCollow, J. 1996. The role of the university and the nature of academic work. *Discourse: Studies in the Cultural Politics of Education* 17 (2): 279-88.

McCormick, J. 1995. *The global environmental movement,* 2nd ed. Chichester, UK: John Wiley and Sons.

McCright, A., and R. Dunlap. 2003. Defeating Kyoto: The conservative movement's impact on US climate change policy. *Social Forces* 50 (3): 348-73.

McDaniels, T., P. Keen, and H. Dowlatabadi. 2006. Expert judgements regarding risks associated with salmon aquaculture practices in British Columbia. *Journal of Risk Research* 9 (7): 775-800.

McIntosh, R. 1985. *The background of ecology: Concept and theory.* New York: Cambridge University Press.

–. 1987. Pluralism in ecology. *Annual Review of Ecological Systems* 18: 321-41.

McMullin, E. 1983. Values in science. In *Proceedings of the 1982 Biennial Meeting of the Philosophy of Science Association,* vol. 1, ed. P. Asquith and T. Nickles, 3-28. East Lansing, MI: The Philosophy of Science Association.

McNeil, K. 2005. Aboriginal rights, resources development, and the source of the provincial duty to consult in *Haida Nation* and *Taku River. Supreme Court Law Review* 29 (2): 447-60.

Meadows, D., D. Meadows, J. Randers, and W. Behrens. 1972. *The limits to growth: A report by the Club of Rome's project on the predicament of mankind.* New York: Universe Books.

Meissner, D. 2002. Central Coast Aboriginals vow to stop fish farms. *Victoria Times-Colonist,* 20 December, A3.

Mendelsohn, E. 1964. The emergence of science as a profession in nineteenth century Europe. In *The management of scientists,* ed. K. Hill, 3-48. Boston: Beacon Press.

–. 1977. The social construction of scientific knowledge. In *The social production of scientific knowledge,* ed. E. Mendelsohn, P. Weingart, and R. Whitley, 3-26. Boston: Ridel.

Merton, R. 1973 [1942]. The normative structure of science. In *The sociology of science,* ed. R. Merton and N. Storer, 267-78. Chicago: University of Chicago Press.

Michalska, A. 2005. *A study of freshwater aquaculture in Canada.* Ottawa: Friends of the Earth Canada.

Miller, K. 1992. Smoking up a storm: Public relations and advertising in the construction of the cigarette problem. *Journalism Monographs* 136.

Mitchell, H., J. Brackett, and B. Hicks. 2003. Relaxing perspectives: The environmental impact of salmon farms. *Issues,* 24 April. Washington Fish Growers' Association, http://www.wfga.net/issues.php.

Mittelstaedt, M. 2004. Farmed salmon are laced with toxins, study finds. *Globe and Mail,* 9 January, A1.

Moore, P. 2004. *Framed salmon: Farmed salmon, PCBs, activists, and the media.* Report for Positive Aquaculture Awareness.

Morton, A. 2004. Dying of salmon farming. In *A stain upon the sea: West coast salmon farming,* ed. S. Hume et al., 199-237. Madeira Park, BC: Harbour Publishing.

Morton, A., R. Routledge, C. Peet, and A. Ladwig. 2001. Sea lice (*Lepeophtheirus salmonis*) infection rates on juvenile pink and chum salmon in the nearshore marine environment of British Columbia, Canada. *Canadian Journal of Fisheries and Aquatic Sciences* 61: 147-57.

Moylan, T. 2000. *Scraps of the untainted sky: Science fiction, utopia, dystopia.* Boulder, CO: Westview Press.

Myers, N., and J. Kent. 2001. *Perverse subsidies: How tax dollars can undercut the environment and the economy,* 2nd ed. Washington, DC: Island Press.

Nadasdy, P. 1999. The politics of TEK: Power and the "integration" of knowledge. *Arctic Anthropology* 36 (1-2): 1-18.

NBSGA (New Brunswick Salmon Growers Association). 2007. Healthy industry. http://www.nbsga.com.

–. n.d.a. Healthy communities. http://www.nbsga.com.

–. n.d.b. Infectious salmon anemia. http://www.nbsga.com/farmedsalmon.php?view=27.

Neshevich, C. 2005. From sea to store. *Food in Canada* (November/December). http://www.bizlink.com/foodfiles/PDFs/dec2005/food_sea2store_dec05.pdf.

Networks of Centres of Excellence Program. 2008. Welcome to the Networks of Centres of Excellence (NCE) programs. http://www.nce.gc.ca.

Newbold, B. 1998. Problems in search of solutions: Health and Canadian Aboriginals. *Journal of Community Health* 23 (1): 59-74.

Newfoundland and Labrador. 2007. Budget speech, 2007: Aquaculture funding programs. St. John's: Government of Newfoundland and Labrador.

Nova Scotia. 2007. Regional aquaculture development advisory committees: Background information. Ministry of Fisheries and Aquaculture, http://www.gov.ns.ca/fish/aquaculture/radac.

Nowotny, H., P. Scott, and M. Gibbons. 2001. *Re-thinking science: Knowledge and the public in an age of uncertainty.* Cambridge, UK: Polity Press.

National Research Council. 2005. Marine and ocean industry: Technology roadmap. http://www.nrc-cnrc.gc.ca.

Nye, M. 1996. *Before big science: The pursuit of modern chemistry and physics 1800-1940.* New York: Twayne.

OCAD (Office of the Commissioner for Aquaculture Development). 2002a. Federal programs and services for five resource-based industries. Report by Paul MacNeil Consulting, December.

–. 2002b. Review of provincial and territorial programs and services in the aquaculture sector. Report by Hewat Consulting, October.

–. 2003. *Achieving the vision.* Ottawa: OCAD.

OECD (Organisation for Economic Co-operation and Development). 2003. Liberalising fisheries markets: Scope and effects. Paris: OECD.

Ommer, R. 1999. Rosie's Cove: Settlement morphology, history, economy, and culture in a Newfoundland outport. In *Fishing places, fishing people,* ed. D. Newell and R. Ommer, 17-31. Toronto: University of Toronto Press.

–. 2007. *Coasts under stress: Restructuring and social-ecological health.* Montreal and Kingston: McGill-Queen's University Press.

O'Riordan, T., and J. Cameron. 1994. *Interpreting the precautionary principle.* London: Earthscan.

PAA (Positive Aquaculture Awareness). 2008. The environment. http://www.farmfreshsalmon.org/F33.cfm.

Palmer, J., and D. Cooper. 1998. *Spirit of the environment: Religion, value, and environmental concern.* New York: Routledge.

Parsons, R., and G. Prest. 2003. Aboriginal forestry in Canada. *Forestry Chronicle* 79 (4): 779-84.

Pauly, D., V. Christenson, R. Froese, and M. Palomares. 2000. Fishing down aquatic food webs. *American Scientist* 88 (1): 46.

Pearce, D., and D. von Finckenstein. 2002. Advancing subsidy reform: Towards a viable policy package. In *Finance for Sustainable Development. Proceedings of the Fifth Expert Group Meeting on Finance for Sustainable Development,* 181-90. New York: United Nations Publications.

Peck, J., and A. Tickell. 1995. The social regulation of uneven development: "Regulatory deficit," England's south east and the collapse of Thatcherism. *Environment and Planning A* 27: 15-40.

–. 2002. Neoliberalizing space. *Antipode* 34 (3): 380-404.

Perreault, T., and P. Martin. 2005. Geographies of neoliberalism in Latin America. *Environment and Planning A* 37: 191-207.

Peters, H. 1995. The interaction of journalists and scientific experts: Co-operation and conflict between two professional cultures. *Media, Culture and Society* 17: 31-48.

Peterson, D., A. Wood, and J. Gardner. 2005. An assessment of Fisheries and Oceans Canada Pacific Region's effectiveness in meeting its conservation mandate. Prepared for the David Suzuki Foundation, Vancouver.

Peterson, K. 2004. "I take this as genocide": Secwepemc battle BC government over land, Aboriginal title. *The Dominion,* 30 September. http://www.dominionpaper.ca.

Petts, J. 2008. Public engagement to build trust: False hopes? *Journal of Risk Research* 11 (6): 821-35.

Phyne, J. 1994. The legal context of industrial aquaculture: Property rights, user conflicts and dispute resolution in three jurisdictions. In *Cooperation in the coastal zone,* ed. P. Wells and P. Ricketts, 405-18. Dartmouth, NS: Coastal Zone Canada Association.

–. 1996. Along the coast and in the state: Aquaculture politics in Nova Scotia and New Brunswick. In *Aquacultural development: Social dimensions of an emerging industry,* ed. C. Bailey, S. Jentoft, and P. Sinclair, 69-92. Boulder, CO: Westview Press.

Phyne, J., and J. Mansilla. 2003. Forging linkages in the commodity chain: The case of the Chilean salmon farming industry, 1987-2001. *Sociologia Ruralis* 43 (2): 108-27.

Podolny, J., T. Stuart, and M. Hannan. 1996. Networks, knowledge, and niches: Competition in the worldwide semiconductor industry, 1984-1991. *American Journal of Sociology* 102 (3): 659-89.

Porter, G. 2003. *Protecting wild Atlantic salmon from impacts of salmon aquaculture.* Washington, DC: World Wildlife Fund and Atlantic Salmon Federation.

Poulantzas, N. 1971. *Pouvoir politique et classes socials.* Paris: Maspero.

Powell, D., and W. Leiss. 1997. *Mad cows and mother's milk: The perils of poor risk communication.* Montreal and Kingston: McGill-Queen's University Press.

Price, D. 1963. *Little science, big science.* New York: Columbia University Press.

PriceWaterhouseCoopers. 2003. *A competitiveness survey of the British Columbia salmon farming industry.* Victoria: Ministry of Agriculture, Food and Fisheries.

Pure Salmon Campaign. n.d. Environmental damage from escaped farm salmon. http://www.puresalmon.org/pdfs/escapes.pdf.

Putnam, R. 1993. *Making democracy work: Civic traditions in modern Italy.* Princeton, NJ: Princeton University Press.

–. 2000. *Bowling alone: The collapse and revival of American community.* New York: Touchstone.

Rabinovich, S. 2007. Can fish farming save depleted cod? *Reuters.com,* 6 February 2007.

Ragosta, J. 1990. Natural resource subsidies and the Free Trade Agreement: Economic justice and the need for subsidy discipline. *George Washington Journal of International Law and Economy* 24: 255-303.

Ramsay, H. 2005. Revolution on Haida Gwaii. *The Tyee,* 21 June. http://thetyee.ca/News/2005/06/21/RevolutionHaida.

Rayner, J., and M. Howlett. 2008. Caught in a staples vise: The political economy of Canadian aquaculture. In *Canada's resource economy in transition,* ed. M. Howlett and K. Brownsey, 121-42. Toronto: Emond Montgomery.

Reber, B., and B. Berger. 2005. Framing analysis of activist rhetoric: How the Sierra Club succeeds or fails at creating salient messages. *Public Relations Review* 31 (2): 185-95.

Reingold, N. 1976. Definitions and speculations: The professionalization of science in the nineteenth century. In *The pursuit of knowledge in the early American republic,* ed. A. Oleson and S. Brown, 33-39. Baltimore: The Johns Hopkins University Press.

Ricardo, D. 1969 [1817]. *The principles of political economy and taxation.* London: Dent.

Riley, S. 2003. Something fishy. *The City – The Coast* 11, no. 3 (4-11 December).

Robson, P. 2006. *Salmon farming: The whole story.* Surrey, BC: Heritage House.

Robinson, E., Y. Gebre, J. Pickering, B. Petawabano, B. Superville, and C. Lavellee. 1995. Effect of bush living on Aboriginal Canadians of the eastern James Bay region with non-insulin-dependent diabetes mellitus. *Chronic Diseases in Canada* 16 (4): 144-48.

Rolston, D., and B. Proctor. 2003. *A baseline report of the incidence of sea lice on juvenile salmonids on British Columbia's north coast*. Vancouver: David Suzuki Foundation.

Roll-Hansen, N. 1994. Science, politics, and the mass media: On biased communication of environmental issues. *Science, Technology and Human Values* 19 (3): 324-41.

Rooney, P. 1992. On values in science: Is the epistemic/non-epistemic distinction useful? In *Proceedings of the 1992 biennial meeting of the Philosophy of Science Association*, vol. 1, ed. D. Hull, M. Forbes, and K. Okruhlik, 13-22. East Lansing, MI: The Philosophy of Science Association.

Rootes, C., ed. 1999. *Environmental movements: Local, national and global*. London: Frank Cass Publishers.

Rose, H. 1983. Hand, brain, heart: A feminist epistemology for the natural sciences. *Signs* 9 (1): 73-90.

Rose, N. 1999. *Powers of freedom: Reframing political thought*. New York: Cambridge University Press.

Rowe, G., and L. Frewer. 2000. Public participation methods: A framework for evaluation. *Science, Technology and Human Values* 25 (3): 3-29.

Rusciano, F. 2003. Framing world opinion in the elite press. In *Framing terrorism: The news media, the government and the public*, ed. P. Norrise, M. Kern, and M. Just, 159-80. New York: Routledge.

Russell, N. 2007. Science and scientists in Victorian and Edwardian literary novels: Insights into the emergence of a new profession. *Public Understanding of Science* 16: 205-22.

Salmon Farm Monitor. 2005. Fishermen call for boycott of New Brunswick farmed salmon. http://www.salmonfarmmonitor.org/intlnewsaugust2005.shtml.

Safarian, A. 1973. Perspectives on foreign direct investment from the viewpoint of a capital receiving country. *Journal of Finance* 28 (2): 419-38.

Salmon, R. 2004. Shellfish marketing: Trends and opportunities. Blue Revolution Consulting Group, Inc. http://www.bluerevolution.ca.

Salmon, R., and B. Kingzett. 2002. Profile and potential of the BC shellfish aquaculture industry. Nanaimo, BC: Kingzett Professional Services.

Sandberg, L., and P. Clancy. 2000. *Against the grain: Foresters and politics in Nova Scotia*. Vancouver: UBC Press.

Sanders, D. 1990. The Supreme Court of Canada and the "legal and political struggle" over indigenous rights. *Canadian Ethnic Studies* 22 (3): 122-29.

SAR (Salmon Aquaculture Review). 1997a. *Salmon aquaculture review: Summary*. Victoria: Government of British Columbia, Environmental Assessment Office.

–. 1997b. *Salmon aquaculture review: First Nations perspectives*, vol. 2. Victoria: Government of British Columbia, Environmental Assessment Office.

Savoie, D. 1992. *Regional economic development: Canada's search for solutions*, 2nd ed. Toronto: University of Toronto Press

Scarfe, B. 2000. After *Delgamuukw*: valuing resource tenures in British Columbia. *Canadian Business Economics* (February): 47-58.

Schabel, U. 2003. God's formula and devil's contribution: Science in the press. *Public Understanding of Science* 12: 255-59.

Schaffer, D. 1995. Shocking secrets revealed! The language of tabloid headlines. *Review of General Semantics* 52: 27-46.

Schreiber, D. 2002. Our wealth sits on the table: Food, resistance, and salmon farming in two First Nations communities. *American Indian Quarterly* 26 (3): 360-77.

Schreiber, D., and D. Newell. 2006a. Negotiating TEK in BC salmon farming: Learning from each other or managing tradition and eliminating contention? *BC Studies* 150: 79-102.

–. 2006b. Why spend a lot of time dwelling on the past? Understanding resistance to contemporary salmon farming in Kwakwaka'wakw territory. In *Pedagogies of the global*, ed. A. Dirlik, 217-32. Boulder, CO: Paradigm.

SCSA (Special Committee on Sustainable Aquaculture). 2007a. *Final report: Volume one*. Victoria: Government of British Columbia.

–. 2007b. *Final report: Volume two*. Victoria: Government of British Columbia.

Sears, P. 1964. Ecology – a subversive subject. *Bioscience* 14 (7): 11-13.

Sedjo, R. 1999. The potential of high-yield plantation forestry for meeting timber needs. *New Forests* 17 (3): 339-60.

Seigrist, M., T. Earle, and H. Gutscher, eds. 2007. *Trust in cooperative risk management*. London: Earthscan.

Sennett, R. 1998. *The corrosion of character*. New York: Norton.

Shepard, P., and D. McKinly. 1969. *The subversive science: Essays toward an ecology of man*. Boston: Houghton Mifflin.

Shumway, S., C. Davis, R. Downey, R. Karney, J. Krauter, J. Parsons, R. Rheault, et al. 2003. Shellfish aquaculture – in praise of sustainable communities and environments. *World Aquaculture* 34 (4): 15-17.

Sider, G. 2003. *Between history and tomorrow: Making and breaking everyday life in rural Newfoundland*. Peterborough, ON: Broadview.

Sierra Legal Defence. 2003. BC First Nations sue fish farms, province, feds. Press release, 22 April.

Simpson, S. 2005. Television's *Boston Legal* takes on BC fish farms. *Vancouver Sun*, 29 September, C1.

Simpson, S., and J. Beatty. 2002. BC to lift moratorium on salmon farming. *Vancouver Sun*, 1 February, A1.

Sinclair, D. 1997. Self-regulation versus command and control? Beyond false dichotomies. *Law and Policy* 19 (4): 529-59.

Sinclair, P., and R. Ommer, eds. 2006. *Power and restructuring: Canada's coastal society and environment*. St. John's: ISER Books.

Sinclair, P., H. Squires, and L. Downton. 1999. A future without fish? In *Fishing people, fishing places*, ed. D. Newell and R. Ommer, 321-39. Toronto: University of Toronto Press.

Slovic, P. 1999. Trust, emotion, sex, politics, and science: Surveying the risk-assessment battlefield. *Risk Analysis* 19 (4): 689-701.

–. 2000. *The perception of risk*. London: Earthscan.

Smith, D. 1987. *The everyday world as problematic: A feminist sociology*. Boston: Northeastern University Press.

Smith, D., J. Vissage, D. Darr, and R. Sheffield. 2001. *Forest resources of the United States, 1997*. Washington, DC: US Department of Agriculture.

Snyder, L., and L. Foster. 1983. An anniversary review and critique: The Tylenol crisis. *Public Relations Review* 9: 24-34.

Sokal, A., and J. Bricmont. 1998. *Fashionable nonsense: Postmodern intellectuals' abuse of science*. New York: Picador.

Sousa-Poza, A. 2000. Well-being at work: A cross-national analysis of the levels and determinants of job satisfaction. *Journal of Socio-Economics* 29 (6): 517-38.

Southwood, R., and P. Henderson. 2000. *Ecological methods*. Malden, MA: Blackwell.

Sparrow, B. 1992. *Uncertain guardians: The news media as a political institution*. Baltimore: The Johns Hopkins University Press.

Staniford, D. 2004. Silent spring of the sea. In *A stain upon the sea: West coast salmon farming*, ed. S. Hume et al., 145-98. Madeira Park, BC: Harbour Publishing.

Statistics Canada. 2003. Census of population: Earnings, levels of schooling, field of study and school attendance. *The Daily*, 11 March.

–. 2005. University enrolments by program and level of instruction. http://www40.statcan.ca.

–. 2006. *Canadian Economic Observer*, March 2006. Catalogue no. 11-010-XIB. Ottawa: Statistics Canada.

–. 2008. *Aquaculture statistics: 2007*. Catalogue no. 23-222-X. Ottawa: Statistics Canada.

Stefanik, L. 2001. Baby Stumpy and the war in the woods: Competing frames of British Columbia forests. *BC Studies* 130: 41-68.

Stickney, R., and J. McVey. 2002. *Responsible marine aquaculture*. Cambridge, MA: CABI Publishing.

Stocking, H. 1999. How journalists deal with scientific uncertainty. In *Communicating uncertainty: Media coverage of new and controversial science,* ed. S. Friedman, S. Dunwoody, and C. Rogers, 23-41. Mahwah, NJ: Lawrence Ehrlbaum Associates.

Stolle, D. 1998. Bowling together, bowling alone: The development of generalized trust in voluntary associations. *Political Psychology* 19 (3): 497-525.

StratCom. 2007. BC polling on closed containment. Coastal Alliance for Aquaculture Reform. http://www.farmedanddangerous.org.

Strydom, P. 2008. Risk communication: World creation through collective learning under complex contingent conditions. *Journal of Risk Research* 11 (1-2): 5-22.

Sturgis, P., and N. Allum. 2004. Science in society: Re-evaluating the deficit model of public attitudes. *Public Understanding of Science* 13 (1): 55-74.

Suzuki, D., and H. Dressel. 1999. *From naked ape to superspecies: Humanity and the global eco-crisis,* revised ed. Vancouver: Greystone Books.

Tank, S. 2004. *Seas of change: Ten recommendations for sustainable fisheries on the BC coast.* Vancouver: David Suzuki Foundation.

Teeple, G. 1995. *Globalization and the decline of social reform.* Toronto: Garamond.

Tennant, P. 1990. *Aboriginal peoples and politics.* Vancouver: UBC Press.

Tesh, S. 1999. Citizen experts in environmental risk. *Policy Sciences* 32 (1): 39-58.

Tollefson, C., and R. Scott. 2006. Charting a course: Shellfish aquaculture and indigenous rights in New Zealand and British Columbia. *BC Studies* 150: 3-41.

Tookenay, V. 1996. Improving the health status of Aboriginal people in Canada: New directions and responsibilities. *Canadian Medical Association Journal* 155: 1581-83.

Trewavas, A. 2001. Urban myths of organic farming. *Nature* 410: 409-10.

Trumbo, C. 1996. Constructing climate change: Claims and frames in US news coverage of an environmental issue. *Public Understanding of Science* 5: 269-83.

UBCIC (Union of British Columbia Indian Chiefs). 1998. Fish farms, zero tolerance: Indian salmon don't do drugs. Open letter to Hon. David Anderson, Minister of Fisheries and Oceans, 8 May.

Ungar, S. 2000. Knowledge, ignorance and the popular culture: Climate change versus the ozone hole. *Public Understanding of Science* 9 (3): 297-312.

Urry, J. 2007. *Mobilities.* Malden, MA: Polity Press.

Vallone, R., L. Ross, and M. Lepper. 1985. The hostile media phenomenon: Biased perception and perceptions of media bias in coverage of the Beirut massacre. *Journal of Personality and Social Psychology* 49: 577-85.

Veenstra, G. 2000. Social capital, SES, and health: An individual-level analysis. *Social Science and Medicine* 50: 619-29.

Vodden, K., and B. Kuecks. 2003. *Clayoquot green economic opportunities project.* Tofino, BC: Friends of Clayoquot Sound.

Vold, T. 2003. Experience developing a results-based forest practices code for British Columbia, Canada. In *Proceedings of the 12th World Forestry Congress.* Quebec City. http://www.gov.bc.ca/for.

Volpe, J. 2001. *Super-unnatural: Atlantic salmon in BC waters.* Vancouver: David Suzuki Foundation. http://www.davidsuzuki.org/files/Super_Un_natural.pdf.

Volpe, J., E. Taylor, D. Rimmer, and B. Glickman. 2000. Evidence of natural reproduction of aquaculture-escaped Atlantic salmon in a coastal British Columbia river. *Conservation Biology* 14 (3): 899-903.

Volpe, J., B. Glickman, and B. Anholt. 2001a. Reproduction of Atlantic salmon (*Salmo salar*) in a controlled stream channel on Vancouver Island, British Columbia. *Transactions of the American Fisheries Society* 130: 489-94.

Volpe, J., B. Anholt, and B. Glickman. 2001b. Competition among juvenile Atlantic salmon (*Salmo salar*) and steelhead trout (*Oncorhynchus mykiss*): Relevance to invasion potential in British Columbia. *Canadian Journal of Fisheries and Aquatic Sciences* 58: 1-11.

Waldram, J., A. Herring, and K. Young. 2006. *Aboriginal health in Canada,* 2nd ed. Toronto: University of Toronto Press.

Walker, K.J., and K. Crowley, eds. 1999. *Australian environmental policy 2: Studies in decline and devolution.* Sydney: University of New South Wales Press.

Wallace, I. 2002. *A geography of the Canadian economy.* Don Mills, ON: Oxford University Press.

Ward, B., and R. Dubos. 1972. *Only one earth.* New York: Deutsch.

Warry, W. 1998. *Unfinished dreams: Community healing and the reality of Aboriginal self-government.* Toronto: University of Toronto Press.

Watkins, M. 1977. The staples theory revisited. *Journal of Canadian Studies* 12: 83-94.

–. 1982. The Innis tradition in Canadian political economy. *Canadian Journal of Political Science and Social Theory* 6 (1-2): 12-34.

Weber, M. 1949 [1919]. Science as a vocation. In *From Max Weber,* ed. H. Gerth and C. Mills, 129-56. New York: Oxford University Press.

WED. 2000. New round of CEAI funding for coastal community projects, hard-hit communities receive the greatest share. News release, 27 January.

–. 2007. Evaluation of the Western Economic Diversification program, Appendix B. Ottawa: Industry Canada.

Weigert, R. 1988. Holism and reductionism in ecology. *Oikos* 53: 267-69.

Weinberg, A. 1961. The impact of large-scale science. *Science* 134: 161-64.

West Coast Environmental Law. 2002. Backgrounder: Bill 74: The Forest and Range Practices Act. http://www.wcel.org.

Westcott, J., K. Hammel, and J. Burka. 2004. Sea lice treatments, management practices and sea lice sampling methods on Atlantic salmon in the Bay of Fundy, New Brunswick, Canada. *Aquaculture Research* 35 (8): 784-92.

Williams, M. 1990. Ecology in an advocate's age. *New Zealand Journal of Ecology* 13: 1-7.

Wilson, D. 1988. Holism and reductionism in evolutionary ecology. *Oikos* 53: 269-73.

Wisenthal, M. 2003. Section W: Education. In *Historical statistics of Canada.* Catalogue no. No. 11-516-XIE. Statistics Canada, http://www.statcan.gc.ca.

Woolford, A. 2005. *Between justice and certainty: Treaty making in British Columbia.* Vancouver: UBC Press.

World Wildlife Fund. 2004. Salmon aquaculture dialogue. http://www.worldwildlife.org/cci/dialogues/salmon.cfm.

Worster, D. 1994. *Nature's economy: A history of ecological ideas,* 2nd ed. New York: Cambridge University Press.

Worm B., E.B. Barbier, N. Beaumont, J. Duffy, C. Folke, B. Halpern, J.B.C. Jackson, et al. 2006. Impacts of biodiversity loss on ocean ecosystem services. *Science* 314: 787-90.

Wroblewski, J., J. Volpe, and D. Bavington. 2006. Manufacturing fish: Transition from wild harvest to aquaculture. In *Power and restructuring: Canada's coastal society and environment,* ed. R. Ommer and P. Sinclair, 145-60. St. John's: ISER Books.

Wurts, W. 2000. Sustainable aquaculture in the twenty-first century. *Reviews in Fisheries Science* 8 (2): 141-50.

Wynne, B. 1993. Public uptake of science: A case for institutional reflexivity. *Public Understanding of Science* 2: 321-37.

–. 1996. May the sheep safely graze? A reflexive view of the expert-lay knowledge divide. In *Risk, environment, and modernity,* ed. S. Lash, B. Szerszinski, and B. Wynne, 45-80. Newbury Park, CA: Sage.

–. 2001. Creating public alienation: Expert cultures of risk and ethics on GMOs. *Science as Culture* 10 (4): 445-81.

Yamagishi, T. 2001. Trust as a form of social intelligence. In *Trust in society,* ed. K. Cook, 121-47. New York: Russell Sage Foundation.

Yearley, S. 2000. Making systematic sense of public discontents with expert knowledge: Two analytical approaches and a case study. *Public Understanding of Science* 9: 105-22.

–. 2003. Social movements as problematic agents of global environmental change. In *Globalization, globalism, environments, environmentalism,* ed. S. Vertovec and D. Posey, 41-54. New York: Oxford University Press.

–. 2005. *Cultures of environmentalism.* New York: Palgrave Macmillan.

Young, J., and J. Werring. 2006. *The will to protect: Preserving BC's wild salmon habitat.* Vancouver: David Suzuki Foundation.

Young, K., J. Reading, B. Elias, and J. O'Neil. 2000. Type 2 diabetes mellitus in Canada's First Nations: Status of an epidemic in progress. *Canadian Medical Association Journal* 163 (5): 561-66.

Young, N. 2004. Environmental risk and populations "at risk." *BC Studies* 141: 59-79.

–. 2006a. *New economic spaces and practices in coastal British Columbia.* PhD dissertation, University of British Columbia.

–. 2006b. Distance as a hybrid actor in rural economies. *Journal of Rural Studies* 22 (3): 253-66.

–. 2007. A return to the commons? Why the environment still matters in democratic theory. *Re-Public,* January. http://www.re-public.gr/en/?p=110.

–. 2008. Radical neoliberalism in British Columbia: Remaking rural geographies. *Canadian Journal of Sociology* 33 (1): 1-36.

–. 2009. The local economy of Port Hardy, BC: Results from a survey of 181 local firms. Report to the Coastal Communities Project and the District of Port Hardy, British Columbia.

Young, N., and R. Matthews. 2003. The social psychology of science in a risk controversy. Paper presented at the Canadian Sociology and Anthropology Association annual meeting, Halifax, 1-4 June.

–. 2007a. Experts' understanding of the public: Knowledge control in a risk controversy. *Public Understanding of Science* 16: 145-62.

–. 2007b. Resource economies and neoliberal experimentation: The reform of industry and community in rural British Columbia. *Area* 39 (2): 176-85.

Zacharias, Y. 2005. Protesting fish farm workers drown out James' speech. *Vancouver Sun,* 27 April, A4.

Index